matscience
symposia
on
theoretical
physics

1

Contributors to this volume:

V. Devanathan
B. Maglić
R. E. Marshak
T. K. Radha
G. Ramachandran
A. Ramakrishnan
K. Raman
T. S. Santhanam
E. Segre
G. Takeda
R. K. Umerjee
K. Venkatesan

matscience
symposia
on
theoretical
physics

Lectures presented at the
1963 First Anniversary Symposium
of the Institute
of Mathematical Sciences
Madras, India

1

Edited by
ALLADI RAMAKRISHNAN
Director of the Institute

 SPRINGER SCIENCE+BUSINESS MEDIA, LLC

ISBN 978-1-4684-7751-1   ISBN 978-1-4684-7749-8 (eBook)
DOI 10.1007/978-1-4684-7749-8

Library of Congress Catalog Card Number 65-21184

©1966 Springer Science+Business Media New York
Originally published by Plenum Press in 1966
Softcover reprint of the hardcover 1st edition 1966

Preface

The MATSCIENCE Institute holds two scientific meetings a year, an anniversary symposium in January to commemorate its birth in 1962 and a three-week summer school in August. The proceedings of the first three meetings were initially made available for private circulation as cyclostyled notes. Professor Rosenfeld, the editor of *Nuclear Physics*, expressed the view that such material, which represented the cooperative effort of the scientists from various countries who participated in the visiting program of our Institute, should be published in a "more permanent form" to reach a wider community of readers.

We were given the opportunity to do this by Mr. Earl Coleman, President of Plenum Press, who made the spontaneous offer, during his visit to Madras just a year ago to publish these proceedings as a continuing series. It was also decided to include in each volume certain lectures delivered during the year, though not at the meeting itself, if they were relevant to the subject matter of the symposium.

The handsome effort of Plenum Press to bring out the series beginning with the very first symposium has been matched by the willing cooperation of our visiting scientists, who have made this an international endeavor, the wholesome consequences of which will be felt beyond the domain of science.

In particular, for the preparation of the first volume, our grateful thanks are due to Professors Segre and Takeda and Dr. Maglic for allowing us to include their lectures in the volume, to Professor Zemach and the publishers of the *Physical Review* for permission to summarize and reproduce figures from his paper in that journal, and to Professor A. H. Rosenfeld for the inclusion of the table of elementary particles from a UCRL Report.

Alladi Ramakrishnan

Madras
November 1965

Introduction

This symposium was arranged as a tribute to our distinguished visitor, Professor R. E. Marshak, who graciously accepted the Niels Bohr visiting professorship of our new Institute of Mathematical Sciences. There is a peculiar appropriateness in naming Professor Marshak as the first occupant of this chair; the very conception of this Institute originated during my visit to America in the spring of 1956, when I had the opportunity to watch the proceedings of the Rochester conference. At that time, to us in India, a world such as this, characterized by high endeavor and creative achievement, seemed beyond reach and almost beyond imagination. The fortuitous visit of Professor Niels Bohr to Madras and his sincere interest in the work of our group transformed vision into reality. This assembly is dedicated to that great ideal, expressed in the words of Professor Niels Bohr as the "worldwide cooperation in science which offers such great opportunities for promoting understanding between all peoples."

Were we not animated by this spirit we would not have had the temerity to invite the sponsor of the foremost international conference in physics to participate in a symposium in which only a small group of young theoreticians is taking part. We are aware that any study in high-energy physics would need the active participation of experimental groups from leading laboratories in the world. The effort made by our group is, I presume, all the more courageous since, despite the absence of experimentalists, it attempted to discuss problems which hold the minds of the most gifted scientists in physics today.

The theory of elementary particles is now offering the greatest challenge since the birth of quantum mechanics. The problems are so complex and baffling that even the novice sometimes feels able to discuss them with the savant without fear of being shown his place, for nobody knows whether we are dealing with a composite theory of elementary particles or an elementary theory of composite particles.

It is a situation which can be very encouraging for a young aspirant to a scientific career; there seems to be work to be done by every class of physicist, from the humblest of calculators to the most exalted of creative scientists.

All the talks herein are semiexpository. Here and there are some strains of speculation and criticism which we offer to our distinguished visitor as our first attempts in this field.

Alladi Ramakrishnan

Editor's Note

Three interesting lectures were delivered during 1962 by Dr. Bogdan. Maglić of CERN, Professor E. Segre of Berkeley, and Professor G. Takeda of Japan during their visits to MATSCIENCE, and since the subjects of these lectures were closely related to those of the symposium, we have included them in this volume with the kind permission and cooperation of the authors.

Contents

Contents of Other Volumes

Symmetries and Resonances*

T. K. RADHA

MATSCIENCE
Madras, India

1. INTRODUCTION

The stimulus for my interest in the subject of symmetries and resonances was the series of lectures by Professor Salam at Trieste entitled "On Lie Groups and Lie Algebra."[1] The theory of symmetries of elementary particle interactions has been revived by a completely new stimulus from the work of Gell-Mann[2] and Neéman,[3] who recognized that the known eight baryons could be assumed to belong to the octet representation of the SU_3 group. Once it is realized that it is the natural generalization of the isospin group to a higher dimension, it seems to be quite possible that various elementary particles and resonances can be classified under this scheme and also that very important relations like the Gell-Mann–Okubo[4] formula, etc., can be derived. Therefore, I feel that any attempt to discuss such possibilities at this stage is justified.

This discussion is divided into five parts:

1. The need for higher symmetries.
2. Charge conservation and charge independence from unitary groups.
3. Resonances in models built on (a) SU_3 with the Sakata model, and (b) SU_3 with the octet model.
4. Global symmetry and models built on other groups.
5. Some speculations.

* The lecture is reproduced as it was prepared in January, 1962, before the discovery of the Ω^--particle. Although it was possible to include new developments at the time of publication, these have been deliberately left out so as not to introduce "disharmony" in the presentation of matter which was based on the then-experimental situation.

Group theoretical concepts necessary for an understanding of this lecture are given in the *Appendix*.

Higher Symmetries

We have learned that the number of known resonant states of strongly interacting particles has increased enormously during the last two years. Before one takes up the question of the dynamics of the elementary forces responsible for the structure of these resonant states, one must know all the internal symmetries and the corresponding quantum numbers.

For strong interactions, we know that there are three exactly conserved quantum numbers (1) isospin I, (2) baryon number N, and (3) hypercharge Y or strangeness S. Therefore, when one looks at the mass spectrum of elementary particles, the natural question is, "Is there any higher symmetry (which perhaps is approximate, as suggested by the mass spectrum again) of strong interactions which of course incorporates I-spin, S, and N conservation?" Such an attempt has been made by Gell-Mann and Neéman[2,3] for the "octet model," and by Salam and Ward[5] and others[6,7] for the "Sakata model." It is interesting to note that certain predictions of these theories have been experimentally verified, for example, the existence of an octet of vector mesons. Another parallel view is that of Sakurai,[8] based on the vector theory of strong interactions—the basic idea is that in order to understand the existence of each conserved current, there should be introduced a vector field which interacts with this current and satisfies gauge invariance. The vector particles proposed were:

1. $I = 1(\rho)$ coupled to I-spin current.
2. $I = 0(\omega)$ coupled to hypercharge current.
3. $I = 0(B)$ coupled to baryon current.
4. $I = \frac{1}{2}(K^*?)$ coupled to (partially conserved) strangeness-changing currents.

Charge Conservation

Any unitary transformation is defined as[9]

$$h_\alpha \rightarrow h'_\alpha = U_{\alpha\beta}h_\beta \qquad (\alpha, \beta = 1, \ldots, n) \qquad (1)$$

where h is an n-component complex vector and U satisfies

$$U^+ = U^{-1} \tag{2}$$

If $h^\alpha = h_\alpha^*$, then expression (1) leads to

$$|h_\alpha|^2 = |h_\alpha'|^2 \tag{3}$$

that is, the transformation U leaves $|h|^2$ invariant. Now, conservation of charge follows from invariance of all interactions under the unitary transformation $p \longrightarrow e^{ie\chi}p$ for all charged fields, where p is a one-dimensional vector. Writing $h = x + iy$, we find that $U = e^{ie\chi}$ leaves $x^2 + y^2$ invariant, that is, U corresponds to rotations in two dimensions (R_2).

Charge Independence

Charge independence may be considered as generalization of two-dimensional rotations to three-dimensional rotations in I-spin space, requiring invariance of strong interactions under these rotations These transformations can be related to the two-dimensional unitary transformations U_2 (in complex space), for instance, by defining

$$h_\alpha = \begin{pmatrix} h_1 \\ h_2 \end{pmatrix} = \begin{pmatrix} p \\ n \end{pmatrix} \tag{4}$$

and

$$h^\alpha = \begin{pmatrix} p^* \\ n^* \end{pmatrix} \tag{5}$$

The most general form of U_2 is

$$U_2 = e^{i\chi 1} e^{i\tau \cdot \chi} \tag{6}$$

where 1 is the unit matrix and τ the 2×2 Pauli matrices. Invariance under $e^{i\chi 1}$ leads to baryon-number conservation, while invariance under $e^{i\tau \cdot \chi} = SU_2$ leads to charge independence. For completeness we shall also write the following:

$$\mathrm{Tr}(\tau_i \tau_j) = 2\delta_{ij} \tag{7}$$

$$[\tau_i, \tau_j] = 2ie_{ijk}\tau_k \qquad \{\tau_i, \tau_j\} = 2\delta_{ij}1 \tag{8}$$

The I-spin operator for $I = \frac{1}{2}$ particles has the representation given by

$$I_i = \frac{\tau_i}{2} \tag{9}$$

For $I = 1$,

$$[I_i, I_j] = ie_{ijk}I_k \tag{10}$$

and I_i has elements

$$I_i^{jk} = -ie_{ijk} \tag{11}$$

$(3 \times 3$ matrices).

To get the higher multiplets, e.g., the mesons, we have merely to use the tensor analysis given in the *Appendix* to get the irreducible representations

$$\pi_0^0 = \frac{1}{\sqrt{2}} h_\alpha h^\alpha = \frac{1}{\sqrt{2}}(pp^* + nn^*)$$

$$\vec{\pi} = h_\alpha h^\beta - \tfrac{1}{2}\delta_\alpha^\beta (h^\gamma h_\gamma)$$

$$= \begin{bmatrix} \tfrac{1}{2}(pp^* - nn^*) & pn^* \\ np^* & -\tfrac{1}{2}(pp^* - nn^*) \end{bmatrix} \tag{12}$$

$$= \begin{bmatrix} \dfrac{\pi^0}{\sqrt{2}} & \pi^+ \\ \pi^- & -\dfrac{\pi^0}{\sqrt{2}} \end{bmatrix}$$

Of course, we have identified the components of the basis of an irreducible representation of a group with a set of physical states having the same space–time properties, for instance, spin, relative parity, G-parity, etc., and $p^2 = m^2$ is also the same for all members of the multiplets.

2. UNITARY SYMMETRY

The generalization of charge independence (SU_2) to higher symmetry leads us to the group $U_3 = e^{i1\theta_k} e^{i\lambda_i\theta_i}$, where $i = 1, 2, \ldots, 8$. There are $3 \times 3 - 1 = 8$ traceless Hermitian matrices (similar to τ) given by, for instance, the following:

$$\lambda_1 = \begin{bmatrix} 0 & 1 & 0 \\ 1 & 0 & 0 \\ 0 & 0 & 0 \end{bmatrix} \qquad \lambda_5 = \begin{bmatrix} 0 & 0 & -i \\ 0 & 0 & 0 \\ i & 0 & 0 \end{bmatrix}$$

$$\lambda_2 = \begin{bmatrix} 0 & -i & 0 \\ i & 0 & 0 \\ 0 & 0 & 0 \end{bmatrix} \qquad \lambda_6 = \begin{bmatrix} 0 & 0 & 0 \\ 0 & 0 & 1 \\ 0 & 1 & 0 \end{bmatrix}$$

$$\lambda_3 = \begin{bmatrix} 1 & 0 & 0 \\ 0 & -1 & 0 \\ 0 & 0 & 0 \end{bmatrix} \qquad \lambda_7 = \begin{bmatrix} 0 & 0 & 0 \\ 0 & 0 & -i \\ 0 & i & 0 \end{bmatrix} \tag{13}$$

$$\lambda_4 = \begin{bmatrix} 0 & 0 & 1 \\ 0 & 0 & 0 \\ 1 & 0 & 0 \end{bmatrix} \qquad \lambda_8 = \begin{bmatrix} 1/\sqrt{3} & 0 & 0 \\ 0 & 1/\sqrt{3} & 0 \\ 0 & 0 & 1/\sqrt{3} \end{bmatrix}$$

The λ_i's satisfy the following relations :

$$\text{Tr}(\lambda_i \lambda_j) = 2\delta_{ij} \tag{14}$$

$$[\lambda_i, \lambda_j] = 2i f_{ijk} \lambda_k \tag{15}$$

$$\{\lambda_i, \lambda_j\} = \tfrac{4}{3}\delta_{ij} + 2d_{ijk}\lambda_k \tag{16}$$

f_{ijk} values are real and totally antisymmetric and d_{ijk} values are real and symmetric, that is, the f_{ijk} are odd under permutations of any two indices while the d_{ijk} are even:

ijk	f_{ijk}	ijk	d_{ijk}	ijk	d_{ijk}
123	1	118	$1/\sqrt{3}$	366	$-\tfrac{1}{2}$
147	$\tfrac{1}{2}$	146	$\tfrac{1}{2}$	377	$-\tfrac{1}{2}$
156	$-\tfrac{1}{2}$	157	$\tfrac{1}{2}$	448	$-1/2\sqrt{3}$
246	$\tfrac{1}{2}$	228	$1/\sqrt{3}$	558	$-1/2\sqrt{3}$
257	$\tfrac{1}{2}$	247	$-\tfrac{1}{2}$	668	$-1/2\sqrt{3}$
345	$\tfrac{1}{2}$	256	$\tfrac{1}{2}$		
367	$-\tfrac{1}{2}$	338	$1/\sqrt{3}$	778	$-1/2\sqrt{3}$
458	$\sqrt{3}/2$	344	$\tfrac{1}{2}$	888	$-1/\sqrt{3}$
678	$\sqrt{3}/2$	355	$\tfrac{1}{2}$		

$$\tag{17}$$

The Sakata Model

First let us consider the Sakata model:

$$h_\alpha = \begin{pmatrix} p \\ n \\ \Lambda \end{pmatrix} \qquad h^\alpha = \begin{pmatrix} p^* \\ n^* \\ \Lambda^* \end{pmatrix} \tag{18}$$

These constitute a minimum set of particles necessary for the manifestation of the I, N, and S quantum numbers. Then the U_3 transformation may be identified with

$$N \longrightarrow e^{i\tau \cdot x} N$$

$$\Lambda \longrightarrow e^{i\beta} \Lambda \tag{19}$$

$$h_\alpha \longrightarrow e^{i\alpha} h_\alpha$$

$$n \longrightarrow \Lambda \qquad \text{(also equivalent to } p \longrightarrow \Lambda)$$

The higher multiplets can be found as follows: We have the representation $D^3(1,0)$ for h_α (see *Appendix*). Therefore all mesons have to be obtained from the product

$$D^3(1, 0) \otimes D^{3^*}(0, 1) = D^8(1, 1) + D^1(0, 0) \tag{20}$$

$$[M_\alpha^\beta = h_\alpha h^\beta - \tfrac{1}{3}\delta_\alpha^\beta (h^\gamma h_\gamma) \text{ and } M_\alpha^\alpha] \tag{21}$$

We may identify the representation $D^8(1,1)$ with pseudoscalar mesons or vector mesons. The I-spin contents give the particles as

I	Y	$J = 0$		$J = 1$	
		State	Mass (MeV)	State	Mass (MeV)
0	0	$\pi_0^0 \ (= \eta)$	560	ω	782
1	0	$\vec{\pi}$	140	ρ	750
$\tfrac{1}{2}$	1	K	494	K^*	888
$\tfrac{1}{2}$	-1	K	494	K^*	888

Thus we have

$$
\Pi = \begin{pmatrix} \dfrac{\pi^0}{\sqrt{2}} + \dfrac{\pi_0^0}{\sqrt{6}} & \pi^+ & K^+ \\[2mm] \pi^- & \dfrac{-\pi^0}{\sqrt{2}} + \dfrac{\pi_0^0}{\sqrt{6}} & K^0 \\[2mm] K^- & \bar{K}^0 & \dfrac{-2\pi_0^0}{\sqrt{6}} \end{pmatrix}
$$

Okubo[4] has given the mass formula for any multiplet of SU_3 as

$$
m = a + bY + c\left[I(I+1) - \frac{Y^2}{4}\right] \tag{22}
$$

This leads to the relation†

$$
\frac{3m_\eta^2 + m_\pi^2}{4} = m_K^2 \tag{23}
$$

which gives $m_\eta = 560$ MeV, which is in very good agreement with experiment. The octet of vector mesons is given by

$$
M = \begin{pmatrix} \dfrac{\rho^{(0)}}{\sqrt{2}} + \dfrac{\omega^{(0)}}{\sqrt{6}} & \rho^{(+)} & K^{*(+)} \\[2mm] \rho^{(-)} & -\dfrac{\rho^{(0)}}{\sqrt{6}} + \dfrac{\omega^{(0)}}{\sqrt{6}} & K^{*(0)} \\[2mm] \bar{K}^{*(-)} & \bar{K}^{*(0)} & -\dfrac{2}{\sqrt{6}}\omega^{(0)} \end{pmatrix} \tag{24}
$$

Similarly, we have from the mass (square) relation

$$
m_{K*}^2 = \frac{m_\rho^2 + 3m_\omega^2}{4} \tag{25}
$$

giving $m_K = 760$ MeV, which is not in agreement with the observed mass of the K^*: $m_{K^*} = 888$ MeV. It is interesting to note that we have got all bosons as composite systems of baryons and antibaryons.

---

†R.P.Feynman has suggested that for bosons we have to take the formula to hold for m^2 rather than for m.

Now let us consider higher baryon multiplets in the Sakata model. We have

$$D^3(1, 0) \otimes D^{3^*}(0, 1) \otimes D^3(1, 0)$$

$$= D^{15}(2, 1) + D^{6^*}(0, 2) + D^3(1, 0) + D^3(1, 0) \tag{26}$$

This gives us

d (dimension)	I	Y	Particles	Mass (MeV)	J
3	0	0	Λ (Y^{***})	1115 (1815)	$\frac{1}{2}(\frac{5}{2}+)$
	$\frac{1}{2}$	1	N (N^{***})	940 (1680)	$\frac{1}{2}(\frac{5}{2}+)$
3	0	0	Y_0^{**}	1520	$(\frac{3}{2}-)$
	$\frac{1}{2}$	1	N^{**}	1512	$(\frac{3}{2}-)$
6	0	2	?	?	
	$\frac{1}{2}$	1	$N_{1/2}^*$ (?)	?	
	1	0	Σ (?)	1190	$\frac{1}{2}$
15	$\frac{1}{2}$	1	Ξ (?)	1315	$\frac{3}{2}$ (?)
	$\frac{3}{2}$	1	N^* (?)	1238	$\frac{3}{2}+$
	1	0	Y_1^* (?)	1385	$\frac{3}{2}+(?)$
	0	0	Y_0^* (?)	1405	$\frac{3}{2}+(?)$
	$\frac{1}{2}$	1	?	?	$\frac{3}{2}+(?)$
	1	2	?	?	$\frac{3}{2}(?)$

Thus, we notice that the only place for the Ξ is in the "15" representation and for Σ in the "6" or "15" representation. If it occurs with Y_1^* and N^* in this representation, it should have spin $\frac{3}{2}$ and positive parity. Further, if we take that the Y_0^* (1405) has spin $\frac{3}{2}+$, then we find that we should have a bound K-N system with mass 1150 MeV (using Okubo's mass formula) and an $N_{1/2}^*$ resonance with mass 1268 MeV and spin $\frac{3}{2}+$.

Thus, we find that the Λ and Σ do not belong to the same representation and also, perhaps, that the Ξ should have spin $\frac{3}{2}$. An alternative suggestion is that the fundamental particles are hidden and the "15" representation may include N, Λ, Σ, Ξ, and two more particles.

The Octet Model

In this model, the primitive objects are hidden. They are defined as

$$l_\alpha = \begin{pmatrix} \nu \\ e^- \\ \mu^- \end{pmatrix} \tag{27}$$

and

$$\bar{L}_\alpha = [D^0\, D^+, S^+] \tag{28}$$

where (ν, e) and (D^0, D^+) form a doublet and μ and S are singlets. For this system, we define the unitary spin as

$$F_i = \frac{\lambda_i}{2}$$

with

$$[F_i, F_j] = if_{ijk} F_k \tag{29}$$

Now form $\bar{L}\lambda_i l / \sqrt{2}$ with

$$i = 1, 2, \ldots, 8 \tag{30}$$

This transforms like an unitary octet and the 8×8 matrices connecting these are given by

$$F_i^{jk} = if_{ijk} \qquad i, j, k = 1, \ldots, 8 \tag{31}$$

We form

$$
\begin{aligned}
&\tfrac{1}{2}\bar{L}(\lambda_1 - i\lambda_2)l \sim \Sigma^+ \sim D^+\nu \\[4pt]
&\tfrac{1}{2}\bar{L}(\lambda_1 + i\lambda_2)l \sim \Sigma^- \sim D^0 e^- \\[4pt]
&\tfrac{1}{2}\bar{L}(\lambda_3)l \sim \Sigma^0 \sim \frac{D^0\nu - D^+ e^-}{\sqrt{2}} \\[4pt]
&\tfrac{1}{2}\bar{L}(\lambda_4 - i\lambda_5)l \sim p \sim S^+\nu \\[4pt]
&\tfrac{1}{2}\bar{L}(\lambda_4 + i\lambda_5)l \sim n \sim S^+ e^- \\[4pt]
&\tfrac{1}{2}\bar{L}(\lambda_6 - i\lambda_7)l \sim \Xi^- \sim D^0\mu^- \\[4pt]
&\tfrac{1}{2}\bar{L}(\lambda_6 + i\lambda_7)l \sim \Xi^0 \sim D^+\mu^- \\[4pt]
&\frac{1}{\sqrt{2}}\bar{L}(\lambda_8)l \sim \Lambda^0 \sim \frac{D^0\nu + D^+ e^- - 2S^+\mu^-}{\sqrt{6}}
\end{aligned}
\tag{32}
$$

that is,

$$N = \begin{bmatrix} \dfrac{\Sigma^0}{\sqrt{2}} + \dfrac{\Lambda^0}{\sqrt{6}} & \Sigma^+ & p \\[2ex] \Sigma^- & -\dfrac{\Sigma^0}{\sqrt{2}} + \dfrac{\Lambda^0}{\sqrt{6}} & n \\[2ex] \Xi^- & \Xi^0 & -\dfrac{2}{\sqrt{6}}\Lambda_0 \end{bmatrix} \tag{33}$$

Thus, the eight known baryons form one degenerate supermultiplet with respect to unitary spin. If we introduce the $(\mu - e)$ mass difference the supermultiplet breaks up into the exactly known multiplets.

For example, assume

$$m_{D^0} = m_{D^+} \qquad m_\nu = m_e \tag{34}$$

Then

$$\begin{aligned} m_N &= m_S + m_e \\ m_\Lambda &= \tfrac{1}{3}(m_D + m_e) + \tfrac{2}{3}(m_S + m_\mu) \\ m_\Sigma &= m_D + m_e \\ m_\Xi &= m_D + m_\mu \end{aligned} \tag{35}$$

and we have

$$\frac{m_N + m_\Xi}{2} = \frac{3m_\Lambda + m_\Sigma}{4} \tag{36}$$

which is in very good agreement with experiment. This follows also from the Okubo mass formula. Thus, the baryons belong to the representation $D^8(1, 1)$. To get the mesons we consider

$$8 \otimes 8 = 1 \oplus 8 \oplus 8 \oplus 10 \oplus \overline{10} \oplus 27 \tag{37}$$

The mesons may belong to the representation D^8, as in the case of Sakata model. Then the matrices connecting the Π_j are just the same as those connecting N_j:

$$F_i^{jk} = -if_{ijk}$$

Now, to couple eight mesons invariantly to the eight baryons, say by γ_5, we have

$$H = 2ig\,\bar{N}\gamma_5\theta_i N \Pi_i \tag{38}$$

where θ satisfies the relation

$$[F_i, \theta_j] = if_{ijk}\theta_k \tag{39}$$

The double occurrence of 8 in the splitting shows there are two independent sets of 8×8 matrices θ_i obeying (39). One may be F_i itself. For the other, define $D_i^{jk} = d_{ijk}$. Then the D's also satisfy (39). The physical difference between these two couplings lies in the symmetry under the operation R, which is not a member of the unitary group. The details of these couplings will be dealt with later by Prof. Marshak.

Let us consider the higher baryon multiplets in the octet model. These also belong to

$$8 \oplus 8 = 1 \oplus 8 \oplus 8' \oplus 10 \oplus \overline{10} + 27$$

8 and 8' correspond to D and F type couplings. We can immediately write down the composition of the different states in the product representation in terms of components of N_j and Π_j as

$(8; \ Y = 1, I = \tfrac{1}{2})^+$

$$= \sqrt{\tfrac{3}{20}}(\sqrt{2}\, n\pi^+ + p\pi^0 + \sqrt{2}\, \Sigma^+ K^0 + \Sigma^0 K^+)$$
$$- \tfrac{1}{\sqrt{20}}(p\chi + \Lambda K^+)$$

$(8; \ Y = 0, I = 0)^0$

$$= \sqrt{\tfrac{1}{5}}(\Sigma^- \pi^+ + \Sigma^+ \pi^- + \Sigma^0 \pi^0 - \Lambda\chi)$$
$$- \sqrt{\tfrac{1}{20}}(pK^- + n\bar{K}^0 + \Xi^- K^+ + \Xi^0 K^0)$$

$(8; \ Y = 0, I = 1)^+$

$$= \sqrt{\tfrac{3}{10}}(p\bar{K}^0 + \Xi^0 K^+) + \sqrt{\tfrac{1}{5}}(\Sigma^+ \chi + \Lambda\pi^+)$$

$(8; \ Y = -1, I = \tfrac{1}{2})^-$

$$= \sqrt{\tfrac{3}{20}}(\sqrt{2}\, \Xi^0 \pi^- + \Xi^- \pi^0 + \sqrt{2}\, \Sigma^- \bar{K}^0 + \Sigma^0 K^-)$$
$$- \sqrt{\tfrac{1}{20}}(\Xi^- \chi + \Lambda K^-) \qquad\qquad (40)$$

$(8'; \ Y = 1, I = \tfrac{1}{2})^+$

$$= \sqrt{\tfrac{1}{12}}(\sqrt{2}\, \Sigma^+ K^0 + \Sigma^0 K^+ - \sqrt{2}\, n\pi^+ + p\pi^0)$$
$$+ \tfrac{1}{2}(\Lambda K^+ - p\chi)$$

$(8'; \ Y = 0, I = 0)^0$

$$= \tfrac{1}{2}[pK^- + n\bar{K}^0 - (\Xi^- K^+ + \Xi^0 K^0)]$$

$(8'; \ Y = 0, I = 1)^+$

$$= \sqrt{\tfrac{1}{3}}(\Sigma^0 \pi^+ - \Sigma^+ \pi^0) + \sqrt{\tfrac{1}{6}}(p\bar{K}^0 - \Xi^0 K^+)$$

$(8'; \ Y = -1, I = \tfrac{1}{2})^-$

$$= \sqrt{\tfrac{1}{12}}(\sqrt{2}\, \Sigma^- \bar{K}^0 + \Sigma^0 K^- - \sqrt{2}\, \Xi^0 \pi^- - \Xi^- \pi^0)$$
$$+ \tfrac{1}{2}(\Lambda K^- - \Xi^- \chi)$$

For "27" we have

$$(27; \ Y = 0, I = 2)^{++} = \Sigma^+ \pi^+$$

$$(27; \ Y = 0, I = 1)^+ = \sqrt{\tfrac{3}{10}}(\Sigma^+ \chi + \Lambda \pi^+) - \sqrt{\tfrac{1}{5}}(p\bar{K}^0 + \Xi^0 K^+)$$

$$(27; \ Y = 0, I = 0)^0$$
$$= \sqrt{\tfrac{27}{40}}\Lambda\chi + \sqrt{\tfrac{3}{40}}(pK^- + nK^0 + \Xi^- K^+ + \Xi^0 K^0)$$
$$+ \sqrt{\tfrac{1}{120}}(\Sigma^+ \pi^- + \Sigma^- \pi^+ + \Sigma^0 \pi^0) \tag{41}$$

$$(27; \ Y = 1, I = \tfrac{3}{2})^{++} = \sqrt{\tfrac{1}{2}}(p\pi^+ + \Sigma^+ K^+)$$

$$(27; \ Y = 1, I = \tfrac{1}{2})^+ = \sqrt{\tfrac{9}{20}}(p\chi + \Lambda K^+)$$
$$+ \sqrt{\tfrac{1}{60}}(p\pi^0 + \Sigma^0 \pi^+ + \sqrt{2}\, n\pi^+ + \sqrt{2}\, \Sigma^+ K^0)$$

$$(27; \ Y = 2, I = 1)^{++} = pK^+$$

Note that $(27; \ Y = -2, I = 1)$, $(27; \ Y = -1, I = \tfrac{3}{2})$, etc., can be obtained from these by operation of R.

The "10" and "$\overline{10}$" are given by

$$(10; \ Y = 1, I = \tfrac{3}{2})^{++} = \sqrt{\tfrac{1}{2}}(p\pi^+ - \Sigma^+ K^+)$$

$$(10; \ Y = 0, I = 1)^+$$
$$= \tfrac{1}{2}(\Sigma^+ \chi - \Lambda \pi^+) + \sqrt{\tfrac{1}{6}}(p\bar{K}^0 - \Xi^0 K^+)$$
$$- \sqrt{\tfrac{1}{12}}(\Sigma^0 \pi^+ - \Sigma^+ \pi^0) \tag{42}$$

$$(10; \ Y = -1, I = \tfrac{1}{2})^-$$
$$= \tfrac{1}{2}(\Lambda K^- - \Xi^- \chi)$$
$$+ \sqrt{\tfrac{1}{12}}(\Xi^- \pi^0 - \Sigma^0 K^- + \sqrt{2}\,(\Xi^0 \pi^- - \Sigma^- \bar{K}^0)$$

$$(10; \ Y = -2, I = 0)^- = \sqrt{\tfrac{1}{2}}(\Xi^0 K^- - \Xi^- \bar{K}^0)$$

$$(\overline{10}; \ Y = -1, I = \tfrac{3}{2})^{--} = \sqrt{\tfrac{1}{2}}(\Xi^- \pi^- - \Sigma^- K^-)$$

$$(\overline{10}; \ Y = 0, I = 1)^+$$
$$= \tfrac{1}{2}(\Sigma^+ \chi - \Lambda \pi^+) + \sqrt{\tfrac{1}{6}}(p\bar{K}^0 - \Xi^0 K^+)$$
$$+ \sqrt{\tfrac{1}{12}}(\Sigma^0 \pi^+ - \Sigma^+ \pi^0) \tag{43}$$

$$(\overline{10}; \ Y = 1, I = \tfrac{1}{2})^+$$
$$= \tfrac{1}{2}(\Lambda K^+ + p\chi)$$
$$+ \sqrt{\tfrac{1}{12}}(p\pi^0 + \sqrt{2}\, n\pi^+ - \Sigma^0 K^+ - \sqrt{2}\, \Sigma^+ K^0)$$

$$(\overline{10}; \ Y = 2, I = 0)^+ = \sqrt{\tfrac{1}{2}}(nK^+ + pK^0)$$

Let us consider the states belonging to the "10" representation. These may correspond to the following :[11]

Y_1^*	$I = 1$	$Y = 0$	1385 MeV	$J = \frac{3}{2} +$
N_1^*	$I = \frac{3}{2}$	$Y = 1$	1238 MeV	$J = \frac{3}{2} +$
Ξ^*	$I = \frac{1}{2}$	$Y = -1$	?	$J = \frac{3}{2} + (?)$
Ω	$I = 0$	$Y = -2$	?	$J = \frac{3}{2} +$

It is interesting to note that there is a unique relation $I = 1 + Y/2$ for this representation. Thus, Okubo's mass relation reduces to

$$m = a + bY \tag{44}$$

This gives

$$m_{\Xi^*} = 1532 \text{ MeV}$$

and

$$m_\Omega = 1679 \text{ MeV} \tag{45}$$

The mass of Ξ^* is in very good agreement with experiment while the Ω seems to be a stable particle (with $S = -3$) which cannot decay into $(\Xi + \bar{K})$ (threshold 1820 MeV). This is to be checked experimentally.

If we apply the mass formula for the imaginary masses also, then by using

$$\Gamma(N_{3/2}^*) = 90 \text{ MeV}$$
$$\Gamma(\Omega) = 0 \tag{46}$$

we get

$$\Gamma(Y_1^*) = 60 \text{ MeV}$$
$$\Gamma(\Xi^*) = 30 \text{ MeV} \tag{47}$$

which is in fairly good agreement with experiments. The Ω can decay only by weak interaction into

$$\Omega \rightarrow \Xi + \pi$$
$$\bar{K} + \Lambda \quad \text{or} \quad \Xi^0 + e^- + \bar{\nu}_e, \text{ etc.} \tag{48}$$
$$\bar{K} + \Sigma$$

It is interesting to note that in 1954 Eisenberg[12] predicted a mass of 1615 MeV for the "new hyperon" if it decays into $\Lambda + \bar{K}$. According to equation (42) we find that for N^*, the decay ratio is

$$\frac{N_1^* \rightarrow p\pi^+}{N_1^* \rightarrow \Sigma + K^+} = 1$$

which is not true, since the ΣK^+ channel is kinematically forbidden.

Thus, we find that unitary symmetry is violated because of the huge mass difference between (π and K) and (p and Σ).

Also, we find that

$$\frac{Y_1^* \rightarrow \Sigma + \pi}{Y_1^* \rightarrow \Lambda + \pi} \simeq \frac{2}{3} \frac{p_\Sigma^3}{p_\Lambda^3} \sim 16\% \tag{49}$$

which experimentally is approximately zero.

However, if we invoke R-invariance,[13] $Y_1^* \rightarrow \pi + \Sigma$ vanishes. But this implies that the $\overline{10}$ should also resonate, leading to a Ξ^* resonance with $I = \frac{3}{2}$ and a bound $K + N$ system. The "27"-fold way leads to the following resonances:

I	Y	State	Mass (MeV)
2	0	Y_2^* resonance	1580 (?) (Observed by Dowell et al.[14])
$\frac{1}{2}$	1	ΛK resonance	1690 (?)
0	0	Y_0^*	1405
$\frac{3}{2}$	-1	Ξ^*	
$\frac{3}{2}$	1	$N_{3/2}^*$	1920 (?)
1	0	Y_1^*	1685 (?)
1	2	$K + N$	

and a group of other $\bar{K}$-baryon and χ-baryon resonances. If Y_1^* (1385) belongs to this, then in agreement with experiment $Y_1^* \rightarrow \Sigma\pi$ is automatically forbidden, as can be seen from (41). However, the $K + p$ scattering shows no sign whatsoever of a bump corresponding to the $K + N$ resonance.

We have further equalities among coupling constants for unitary symmetry with R-symmetry (for D-type coupling) :

$$G_{\pi\Lambda\Sigma}^2 = \tfrac{4}{3} G_{\pi NN}^2 \tag{50a}$$

$$G_{\pi NN}^2 = G_{\pi\Xi\Xi}^2 \tag{50b}$$

$$G_{\pi\Sigma\Sigma}^2 = 0 \tag{50c}$$

$$G_{K\Sigma N}^2 = 3G_{K\Lambda N}^2 \tag{50d}$$

$$G_{K\Lambda N}^2 = G_{K\Xi\Lambda}^2 \tag{50e}$$

$$G^2_{K\Sigma N} = G^2_{K\Xi\Sigma} \tag{50f}$$

$$G^2_{K\Sigma N} = G^2_{\pi N N}. \tag{50g}$$

Of these (b), (c), (e), and (f) follow from R-invariance alone. We see from (g) that K- and π-coupling to N is the same, which contradicts the deductions from photoproduction experimental data, from which we find that π-coupling should be much stronger than K-coupling. Perhaps the violation of unitary symmetry can be calculated to give

$$\frac{G^2_{\pi N N}}{G^2_{K\Sigma N}} \sim \frac{m^2_K}{m^2_\pi} \tag{51}$$

The $J = \frac{3}{2}-$ resonances may be due to vector boson–baryon couplings just as PS meson–baryon coupling leads to $P_{3/2^+}$ resonances.

For the vector mesons we may write the decay widths as

$$\Gamma_{K^* \to K\pi} = \frac{2\gamma^2_{K^* K\pi} k^3_1}{4\pi m^2_{K^*}} \tag{52}$$

$$\Gamma_{\rho \to 2\pi} = \frac{8}{3} \frac{\gamma^2_{\rho\pi\pi}}{4\pi} \frac{k^3_2}{m^2_\rho} \tag{53}$$

which for unitary symmetry will give

$$\frac{\Gamma_{K^* \to K\pi}}{\Gamma_{\rho \to 2\pi}} = \frac{3}{4} \frac{k^3_1}{k^3_2} \frac{m^2_\rho}{m^2_{K^*}} \tag{54}$$

Then we get, using $m_{K^*} = 880$ MeV and the ρ mass and width $\Gamma_{K^* \to K\pi}$ to be 30 MeV, while experimentally it is 50 MeV.[15]

Further, we have

$$\frac{f^2_\omega}{4\pi} = \frac{3}{4} \frac{f^2_\rho}{4\pi} \tag{54a}$$

3. GLOBAL SYMMETRY

Global symmetry was perhaps the first attempt at a higher symmetry.[16] It was assumed that the N, Λ, Σ, and Ξ form a supermultiplet symmetrically coupled by very strong π interactions and degenerately but unsymmetrically coupled by moderately strong K interactions. The group theoretical arguments are based on $\mathscr{G}_0$,[17] which has 8×8 unitary representation to which the baryons belong, that is,

$$B = \begin{bmatrix} N \\ \Xi \\ Y \\ Z \end{bmatrix} = \begin{bmatrix} p \\ n \\ \Xi^0 \\ \Xi^- \\ \Sigma^+ \\ Y^0 \\ Z^0 \\ \Sigma^- \end{bmatrix} \tag{55}$$

and

$$\mathscr{G}_0 = U_L \begin{bmatrix} aI & bI & 0 & 0 \\ -b^*I & a^*I & 0 & 0 \\ 0 & 0 & a'I & b'I \\ 0 & 0 & -b'^*I & a'^*I \end{bmatrix} \tag{56}$$

where

$$U_L = e^{i(u_1 U_1 + u_2 U_2 + u_3 U_3)} \tag{57}$$

and U can be L, M, or N (with corresponding small l, m, n) where

$$L_1, L_2, L_3 = L_i = \tfrac{1}{2} \begin{bmatrix} \sigma_i & & & \\ & \sigma_i & & \\ & & \sigma_i & \\ & & & \sigma_i \end{bmatrix}$$

with $i = 1, 2, 3$.

$$M_1 = \tfrac{1}{2} \begin{bmatrix} 0 & 0 & 0 & 0 \\ 0 & 0 & 0 & 0 \\ 0 & 0 & 0 & I \\ 0 & 0 & I & 0 \end{bmatrix} \qquad N_1 = \tfrac{1}{2} \begin{bmatrix} 0 & I & 0 & 0 \\ I & 0 & 0 & 0 \\ 0 & 0 & 0 & 0 \\ 0 & 0 & 0 & 0 \end{bmatrix}$$

$$M_2 = \tfrac{1}{2}i \begin{bmatrix} 0 & 0 & 0 & 0 \\ 0 & 0 & 0 & 0 \\ 0 & 0 & 0 & -I \\ 0 & 0 & I & 0 \end{bmatrix} \qquad N_2 = \tfrac{1}{2}i \begin{bmatrix} 0 & -I & 0 & 0 \\ I & 0 & 0 & 0 \\ 0 & 0 & 0 & 0 \\ 0 & 0 & 0 & 0 \end{bmatrix} \tag{58}$$

$$M_3 = \tfrac{1}{2} \begin{bmatrix} 0 & 0 & 0 & 0 \\ 0 & 0 & 0 & 0 \\ 0 & 0 & I & 0 \\ 0 & 0 & 0 & -I \end{bmatrix} \qquad N_3 = \tfrac{1}{2} \begin{bmatrix} I & 0 & 0 & 0 \\ 0 & -I & 0 & 0 \\ 0 & 0 & 0 & 0 \\ 0 & 0 & 0 & 0 \end{bmatrix}$$

One gets

Resonance	Representation	Total energy (MeV)	Decay into	ω	Relative partial width
$(N^*)_{3/2}^{++}$	$(\frac{3}{2}, \frac{1}{2}, 0)$	1237	$\pi^+ + p$	1	1
$(Y_1^*)^+$		1385	$\pi^+ + \Lambda$	$\frac{2}{3}$	0.5
			$\Sigma^+ + \pi^0$	$\frac{1}{6}$	0.03
			$\Sigma^0 + \pi^+$	$\frac{1}{6}$	0.03
$(Z^*)_2^{++}$		1539	$\pi^+ + \Sigma^+$	1	1.9 (?)
$(\Xi^*)_{3/2}^+$		1637	$\pi^+ + \Xi^0$	1	1.5 (?)

where widths are calculated using CG coefficients $\times (q^3 E_B)/(E_B + E_M)$ (q momentum of π in rest frame of resonance) and energy (mass)

$$E^* = E_{N^*} + \alpha' L \cdot M + \beta'(n^2 - \tfrac{3}{4}) + \gamma'(n_3 - \tfrac{1}{2}) \tag{59}$$

$$E = E_N + \alpha L \cdot M + \beta(n^2 - \tfrac{3}{4}) + \gamma(n_3 - \tfrac{1}{2}) \tag{60}$$

This leads to the same old mass relation

$$\frac{3m_\Lambda + m_\Sigma}{4} = \frac{m_N + m_\Xi}{2} \tag{61}$$

C_2 Model

The basis for the five-dimensional representation $D^5(0, 1)$ may be taken as N, Λ, Ξ.[18] This naturally implies that $N. \Lambda, \Xi$ space–time properties are the same, but Σ may be different. For mesons we have

$$D^5(0, 1) \otimes D^5(0, 1) = D^1 \oplus D^{10} \oplus D^{14} \tag{62}$$

(Note that ψ_a and ψ^a are equivalent for C_2). Thus we have the representation shown in Fig. 1; we have in addition to the usual octet of mesons those with $I = 0$ and $Y = 2$ and -2. This is similar for vector mesons (resonances), also. For higher baryon multiplets, we use

$$D^5(0, 1) \otimes D^{10}(2, 0) = 35 \oplus 10 \oplus 5 \tag{63}$$

The lowest representation which can accommodate Σ is D^{10}. This means we should have baryon resonances with $J = \frac{1}{2}$ and the following quantum numbers :

$$I = \tfrac{1}{2} \qquad Y = 1 \qquad \pi N \text{ res.}$$

$$I = \tfrac{1}{2} \qquad Y = -1 \qquad \Xi \pi \text{ res.}$$

$$I = 0 \qquad Y = 2 \qquad NK \text{ res.}$$

$$0 \qquad \bar{K}N + \Xi K \text{ res.}$$

$$-2 \qquad \Xi \bar{K} \text{ res.}$$

Of course, the masses are not prescribed.

B_2 Model

For this, $D^4(1, 0)$ is the basic representation we are interested in, with the basis given by N, Ξ. The Λ belongs to the singlet representation $D^1(0, 0)$. We have for K mesons

$$D^4(1, 0) \otimes D^1(0, 0) = D^4(1, 0) \qquad (64)$$

$D^4(1,0)$ thus gives K^+, K^-, K^0, and $\bar{K}^0$. The π-meson and other baryons are obtained from

$$D^4(1, 0) \otimes D^4(1, 0) = D^{10} \oplus D^5 \oplus D^1 \qquad (65)$$

The "5" representation may include $\Sigma^{\pm,0}$ $(\pi^{\pm 0})$ and two other baryons $X^\pm$ $(D^\pm)$ with $I = 0$, $Y = 0$.

Also, Y_1^* may belong to representation "5." Then $Y_1^* \rightarrow \Sigma + \pi$ is forbidden, since Σ and π each belong to the "5" representation and $5 \otimes 5$ does not contain a "5" again. But $Y_1^* \rightarrow \Lambda + \pi$ is allowed. Notice further that this representation does not require the π mass to be the same as the K mass.

G_2 Model

The basis for the representation $D^7(1, 0)$, may be taken as Σ, N, Ξ, and Λ belongs to $D^1(0, 0)$. This can accommodate odd $\Sigma\Lambda$ parity. Similarly, we have

$$D^7(1, 0) \rightarrow K, \bar{K}, \pi \qquad D^1 \rightarrow \eta \qquad (66)$$

and

$$D^7(1, 0) \rightarrow K^*, \bar{K}^*, \rho \qquad D^1 \rightarrow \omega$$

For baryon resonances we have

$$D^7(1, 0) \otimes D^7(1, 0) = D^{27} \otimes D^{14} \oplus D^7 \oplus D^7 \qquad (67)$$

An $I = \frac{3}{2}$ resonance is allowed only in the "14" and "27" representations. The "14"-fold way requires

$I = \frac{3}{2}$	$Y = 1$	N^*
$I = \frac{3}{2}$	$Y = 1$	Ξ^*
$I = 1$	$Y = 0$	?
$I = 0$	$Y = 2, 0, -2$	?

The $\Lambda\pi$ (Y_1^*) resonance belongs to representation "7." This multiplet does not contain $I = \frac{3}{2}$. But it may include $I = \frac{1}{2}$, $\Xi\pi$ resonance, and $N\pi$ resonances.

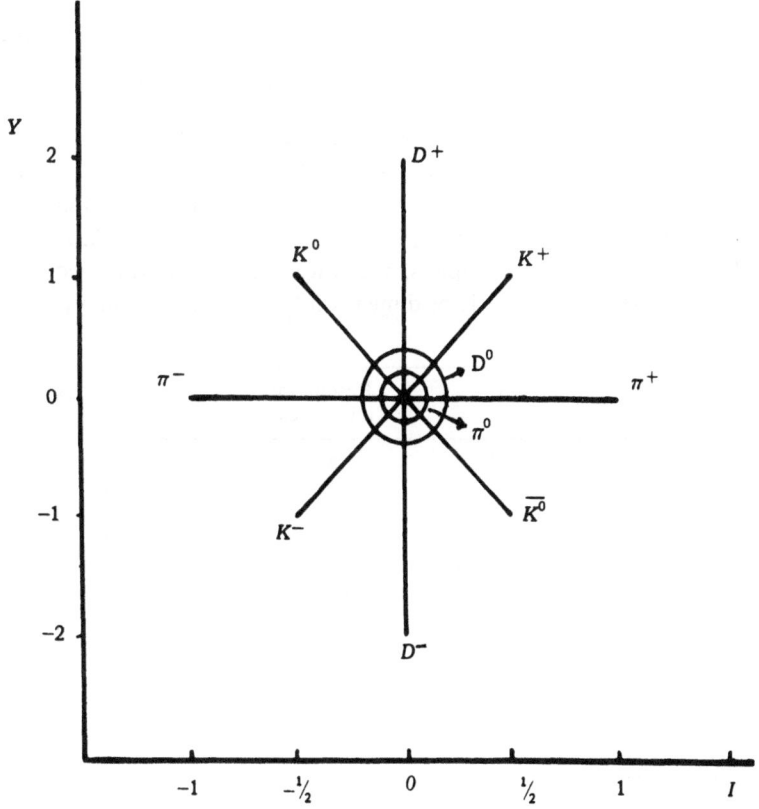

Fig. 1. Mesons in the representation "14" of C_2.

Sawyer[19] has recently considered global symmetry as invariance under the group operation $SU_3 \otimes SU_2$, which is a subgroup of SU_6. He takes the basis as

$$B_6 = \begin{pmatrix} N \\ Y \\ Z \end{pmatrix} = \begin{pmatrix} p \\ n \\ \Lambda \\ \Sigma^+ \\ \Sigma^0 \\ \Sigma^- \end{pmatrix} \tag{68}$$

Thus, SU_3 will act on $\begin{pmatrix} N \\ Y \\ Z \end{pmatrix}$, while SU_2 will act on each of the three: N, Y, and Z. To get the mesons we consider

$$B_6^* \otimes B_6 = 1 \oplus 35 \quad \text{under } SU_6 \tag{69}$$

$$= 1 \oplus 3 \oplus 8 \oplus 24 \quad \text{under } SU_3 \otimes SU_2 \tag{70}$$

The representation "3" gives the pseudoscalar π (and ζ for vector states). Representation "8" consists of K, η, π' (K^*, ω, and ρ) and "24" consists of higher mass bosons. The merit of this group consists in the fact that π belongs to a different representation than K, η, π'.

For higher baryon multiplets, the lowest multiplet which contains an $I = \frac{3}{2}$, the $S = 0$ state is of dimension "12", corresponding to

I	S	Resonance on particle	Mass (m_π)
$\frac{3}{2}$	0	$N_{3/2}^*$	9
1	-1	Y_1^*	10
0	-1	Y_0^*	11
$\frac{1}{2}$	-2	Ξ	9.4
$\frac{1}{2}$	0	$N_{1/2}^*$	?

4. SOME SPECULATIONS

First let us consider Okubo's mass relation

$$m = a + bY + c\left[I(I+1) - \frac{Y^2}{4}\right] \tag{1}$$

Let us apply this to the representation "15" of the higher multiplets of baryons in the Sakata model. The I-spin, Y assignments for these are given by

I.

$$I = \tfrac{1}{2} \qquad Y = 1 \tag{2}$$

$$\left.\begin{array}{ll} I = 0 & Y = 0 \\ I = 1 & Y = 2 \end{array}\right\} \quad I = \frac{Y}{2} \tag{3}$$

$$\tag{4}$$

II.

$$I = \tfrac{3}{2} \qquad Y = 1 \tag{5}$$

$$\left.\begin{array}{ll} I = 1 & Y = 0 \\ I = \tfrac{1}{2} & Y = -1 \end{array}\right\} \quad I = 1 + \frac{Y}{2} \tag{6}$$

$$\tag{7}$$

It is interesting to note that for the first three [(2) to (4)], the relation $I = Y/2$ holds, while for the second three [(5) to (7)], the relation is $I = 1 + Y/2$. Thus, Okubo's mass relation immediately becomes a linear relation for these, with

$$m_I = a + bY \tag{8}$$

and

$$m_{II} = a' + b'Y \tag{9}$$

Let us suppose that Y_1^* (1385) and $N_{3/2}^*$ (1238) belong to the second half, with $J = \tfrac{3}{2} +$. Then we can predict the mass of the third particle of II with $I = \tfrac{1}{2}$, $Y = -1$ as

$$m(Y = -1, I = \tfrac{1}{2}) = 1532 \text{ MeV} \tag{10}$$

which can be easily identified as the Ξ^*-resonance and not as the Ξ-particle. So the natural conclusion would be that if the Y_1^* and $N_{3/2}^*$ belong to the "15" representation, then the Ξ should not be placed in the same multiplet, but may have a place only in a further higher multiplet, say from $3 \times 3^* \times 3 \times 3^* \times 3$; it is quite probable that it will be grouped with particles with spin $\tfrac{1}{2}$. A study of the irreducible representations of $3 \times 3^* \times 3 \times 3^* \times 3$ is being done.

It is worthwhile mentioning that to predict the mass of the third particle in the division I (or II), it is necessary to know only two masses of the same group I (or II) in view of the linear relation between I and Y. Further, knowing the three masses of the same group I (or II) is not sufficient to give the mass of the other group, although we have only three unknowns in equation (I). So if we further assume that the Y_0^* resonance with mass 1405 MeV is also $J = \tfrac{3}{2} +$ then we get

$$m(Y = 2, I = 1) = 1131 \text{ MeV} \tag{11}$$

which should be a bound state of $(K + N)$. Also, we get

$$m_{N^*}(Y = 1, I = \tfrac{1}{2}) = 1265 \text{ MeV} \tag{12}$$

which is yet to be experimentally seen. Alternatively, if we assume Y_0^* (1520 MeV) and $N_{1/2}^*$ (1512 MeV) belong to "15", then all the particles of this group should have spin $J = \tfrac{3}{2} -$. This gives a $K + N$ resonance with mass

$$m = 1504 \text{ MeV} \tag{13}$$

It is interesting to note that $K_{I=1/2}^*$ (730 MeV), $\zeta_{I=1}$ (560 MeV), and $\omega_{I=0}$ (752 MeV) satisfy the mass relation

$$m_{K^*}^2 = \frac{m_\zeta^2 + 3m_\omega^2}{4}$$

very well.

The particle belonging to the "27"-fold way can also be split as follows:

$$
\left.
\begin{array}{ll}
I = 0 & Y = 0 \\
I = \tfrac{1}{2} & Y = 1 \\
I = 1 & Y = 2
\end{array}
\right\} \; I = \dfrac{Y}{2}
$$

$$
\left.
\begin{array}{ll}
I = 2 & Y = 0 \\
I = 1 & Y = -2 \\
I = \tfrac{3}{2} & Y = -1
\end{array}
\right\} \; I = 2 + \dfrac{Y}{2} \tag{14}
$$

$$
\left.
\begin{array}{ll}
I = 1 & Y = 0 \\
I = \tfrac{3}{2} & Y = 1 \\
I = \tfrac{1}{2} & Y = -1
\end{array}
\right\} \; I = 1 + \dfrac{Y}{2}
$$

so that, knowing any two of the same group, the third can be predicted by the linear mass relation.

In the "octet" model, the higher baryon multiplets correspond to

$$8 \otimes 8 = 1 \oplus 8 \oplus 8 \oplus 10 \oplus \overline{10} \oplus 27$$

We have already discussed "10" in section 3. Let us now consider the $\overline{10}$. We know that if Y_1^*, $N_{3/2}^*$ belong to the "10" representation, there should be yet another baryon with mass 1670 MeV and a Ξ^* resonance with mass 1532 MeV. (Notice the similarity to the Sakata model "15"

II; the same relation $I = 1 + Y/2$ holds for both.) However, the decay
widths

$$\frac{\Gamma(Y_1^* \rightarrow \Sigma\pi)}{\Gamma(Y_1^* \rightarrow \Lambda\pi)} \sim 16\% \tag{15}$$

according to unitary symmetry if Y_1^* belongs to "10." We know experimentally it is almost $\sim 0\%$. This can be overcome by imposing R
invariance. Now let us study the further consequences of R symmetry.
As pointed out by Sakurai and Glashow, if the "10" resonates, so should
the $\overline{10}$. This leads to the following resonances :

1. Ξ^* resonance with $I = \frac{3}{2}$ $(RN_{3/2}^*)$

2. $N_{1/2}^*$ resonance with $I = \frac{1}{2}$ $(R\Xi_{1/2}^*)$

                                                               (16)

3. Y_1^* resonance with $I = 1$ (RY_1^*)

4. $\chi\,(?)$ resonance with $I = 0$ $(R\Omega)$

We may predict the masses of two of these particles knowing the
other two. We do not have experimental information on any of these,
except perhaps the Y_1^*. But we may assume (using the effect of R-
operation) that the masses of $\Xi_{3/2}^*(\Xi_{1/2}^*)$ and $N_{3/2}^*(N_{1/2}^*)$ transform in the
same way as Ξ and N. This gives

$$m_{\Xi_{3/2}^*} \simeq 1735 \text{ MeV} \tag{17}$$

$$m_{N_{1/2}^*} \simeq 1094 \text{ MeV} \tag{18}$$

Using this we get

$$m_{Y_1^*} \simeq 1415 \text{ MeV} \quad \text{and} \quad m_{(K^+N)} \simeq 936 \text{ MeV}$$

which again leads to a bound K^+N system.

In the octet model we know that N, Λ, Σ, and Ξ belong to the "8"
representation of SU_3. Now, the recently observed $N_{1/2}^*$ with mass 1685
MeV and Y_0^* with mass 1815 MeV are supposed to be on the same
Regge trajectory as the N and Λ, respectively (if their spin is $J = \frac{5}{2} +$),
with slope given by $\sim \frac{1}{50} m_\pi^2$. Let us calculate the mass of the Y_1^* and
Ξ^* with $J = \frac{5}{2} +$, which will be on the Σ and Ξ Regge trajectories,
respectively.

Table I. Group SU_3

Representation	d	I-spin content	$\otimes D^3 (1, 0)$	$\otimes D^6 (2, 0)$	$\otimes D^8 (1, 1)$
$D(0, 0)$	1	0	3	6	8
$D(1, 0)$	3	$0, \frac{1}{2}$	$6 + 3^*$	$10 + 8$	$15 + 6^* + 3$
$D(0, 1)$	3^*	$0, \frac{1}{2}$	$8 + 1$	$15 + 3$	$15^* + 6 + 3^*$
$D(2, 0)$	6	$0, \frac{1}{2}, 1$	$10 + 8$	$15' + 15 + 6^*$	$24 + 15^* + 6 + 3^*$
$D(1, 1)$	8	$0, \frac{1}{2}, \frac{1}{2}, 1$	$15 + 6^* + 3$	$24 + 15^* + 6 + 3^*$	$27 + 10 + 10^* + 8$ $+ 8 + 1$
$D(0, 3)$	10^*	$0, \frac{1}{2}, 1, \frac{3}{2}$			
$D(3, 0)$	10	$0, \frac{1}{2}, 1, \frac{3}{2}$			
$D(2, 1)$	15	$0, \frac{1}{2}, \frac{1}{2}, 1, 1, \frac{3}{2}$			
$D(2, 2)$	27	$0, \frac{1}{2}, \frac{1}{2}, 1, 1, 1, \frac{3}{2}, \frac{3}{2}, 2$			

Table II. Group $C_2[B_2]$

Representation	d	I-Spin content	$\otimes D^4\,(1,0)$	$\otimes D^5\,(0,1)$	$\otimes D^{10}\,(2,0)$
$D(0,0)$	1	0	4	5	10
$D(1,0)$	4	$0, 0, \frac{1}{2}$ $[\frac{1}{2}, \frac{1}{2}]$	$10+5+1$	$16+4$	$40+16$
$D(0,1)$	5	$0, \frac{1}{2}, \frac{1}{2}$ $[0,0,1]$	$16+4$	$14+10+1$	$35'+10+5$
$D(2,0)$	10	$0, 0, 0, \frac{1}{2}, \frac{1}{2}, 1$ $[0, 1, 1, 1]$	$20+16+4$	$35'+10+5$	
$D(0,2)$	14	$0, \frac{1}{2}, \frac{1}{2}, 1, 1, 1$ $[0, 0, 0, 1, 1]$	$40+16$	$35+30+5$	

Table III. Group G_2

Representation	d	I-Spin content	$\otimes D^1\,(1,0)$	$\otimes D^{14}\,(0,1)$
$D(1,0)$	7	$\frac{1}{2}, \frac{1}{2}, 1$	$27+14+7+1$	$64+27+7$
$D(0,1)$	14	$0, 0, 0, 1, \frac{3}{2}$	$64+27+7$	$77+77'+27+14+1$
$D(2,0)$	27	$0, \frac{1}{2}, \frac{1}{2}, 1, 1, 1,$ $\frac{3}{2}, \frac{3}{2}, 2$	$77+64+27$ $+14+7$	

For $d\alpha/dt \sim 1/(50m_\pi^2)$, we get the following masses:

B	Mass	B*	Predicted mass (MeV)	Experimental value (MeV)	
N	940	$N^*_{1/2}$	1685	1688	(19)
Λ	1115	Y^*_0	1793	1815	
Σ	1190	Y^*_1	1836	?	
Ξ	1315	$\Xi^*_{1/2}$	1920	?	

It would be interesting if these belonged to the same representation "8." In this case, using Okubo's mass relation and three of the above masses, the fourth can be predicted. Let us assume the first three:

$$1685 = a + b$$

$$1793 = a$$

$$1836 = a + 2c$$

This gives

$$m_{\Xi^*} = a - b + \frac{c}{2} = 1923 \text{ MeV} \tag{20}$$

which is in agreement with that given in (19). The experimental mass values for $N^*_{1/2}$ and Y^*_0 are fairly in good agreement with the Regge trajectory values. One may thus expect two more $J = \frac{5}{2} +$ resonances at about the energies shown both by symmetry arguments and from Regge trajectories.

We can generalize the above considerations for all Regge trajectories and study in general the consequences of higher symmetries and the hypothesis of Regge poles together. One remark that can be made is that if $d\alpha/dE$ is approximately constant [rather than $d\alpha/dt(= E^2) \sim 1/(50\ m_\pi^2)$] then Okubo's mass relation will hold for higher particles on the Regge trajectory automatically. But if $d\alpha/dt \sim 1/(50\ m_\pi^2)$, then Okubo's mass relation will be satisfied by these higher mass particles in the Regge trajectory only if

$$\frac{m_N^2 + m_\Xi^2}{2} \sim \frac{3m_\Lambda^2 + m_\Sigma^2}{4} \tag{21}$$

which seems to be satisfied for N, Λ, Σ, and Ξ.

APPENDIX

Group theoretical concepts necessary for this discussion are summarized here. We give the definition of the following compact Lie groups as:

1. A_l—order $(l^2 + 2l)$—group of unitary unimodular matrices in complex space of $(l + 1)$ dimensions.
2. B_l—order $(2l^2 + l)$—group of orthogonal transformations (rotations) in a real space of $(2l + 1)$ dimensions.
3. C_l—order $(2l^2 + l)$—group of unitary matrices U in complex space of $2l$ dimensions.
4. G_2—exceptional group of order 14.

Take the case $l = 2$. This implies there are two commuting matrices which can be diagonalized simultaneously and which correspond to the number of conserved quantities. Any irreducible representation of these groups can be labeled by means of two nonnegative integers a_1 and a_2 which are related to Y and I for strong interactions.

We give below in tables the different groups and their dimensions in terms of a_1 and a_2 and also the I-spin content for the different representations and the decomposition of product representations.

Tensor Analysis of SU_3

The irreducible representations into which product representation breaks can be given as follows:

(1) $$\psi_a \otimes \psi_b \sim \psi_{ab,} \oplus \psi_{a,b}$$

where

$$\psi_{ab,} = \frac{\psi_a \psi_b + \psi_b \psi_a}{2}$$

$$\psi_{a,b} = \frac{\psi_a \psi_b - \psi_b \psi_a}{2}$$

(2) $$\psi_a \otimes \psi^b = (\psi_a^b - \tfrac{1}{3}\delta_a^b \psi_c^c) \oplus \delta_a^b \psi_c^c$$

(3) $$\psi_a \otimes \psi_b \otimes \psi^c \sim \psi_{ab,}^c \oplus \psi_{a,b}^c \oplus \psi_{ac}^c \psi_{cb}^c$$

(4) $$\psi_a \otimes \psi_b \otimes \psi_c \sim \psi_{abc,} \oplus \psi_{ab,c} \oplus \psi_{ac,b} \oplus \psi_{a,b,c}$$

where, for instance,

$$\psi_{ab,c} = \tfrac{1}{4}(\psi_{abc} - \psi_{acb} + \psi_{bac} - \psi_{bca})$$

with

$$\psi_{abc} = \psi_a \psi_b \psi_c, \text{ etc.}$$

REFERENCES

1. A. Salam, Lectures delivered at Summer School held at Trieste (1963).
2. M. Gell-Mann, CTSL-20 (1961); *Phys. Rev.* **125**: 1067 (1962).
3. Y. Neéman, *Nucl. Phys.* **26**: 222 (1961).
4. S. Okubo, *Progr. Theoret. Phys.* **27**: 949 (1962).
5. A. Salam and J. Ward, *Nuovo Cimento* **20**: 419 (1961).
6. M. Ikeda, S. Ogawa, and Y. Ohnuki, *Progr. Theoret. Phys.* **22**: 715 (1959).
7. J. Wess, *Nuovo Cimento* **10**: 15 (1960).
8. J. J. Sakurai, *Ann. Phys.* **11**: 1 (1960).
9. P. T. Mathews and A. Salam, *Proc. Phys. Soc. (London)* **80**: 28 (1962).
10. J. J. Sakurai and S.L. Glashow, *Nuovo Cimento* **25**: 337 (1962).
11. J. J. Sakurai and S.L. Glashow, *Nuovo Cimento* **26**: 622 (1962).
12. Y. Eisenberg, *Phys. Rev.* **96**: 541 (1954).
13. J. J. Sakurai, *Phys. Rev. Letters* **7**: 426 (1961).
14. Dowell *et al.*, Proceedings of the Aix-en-Provence Conference on Elementary Particles, Vol. 1, CEN, Saclay, France, 1961, p. 385.
15. A. H. Rosenfeld, UCRL-10492 (1962).
16. M. Gell-Mann, *Phys. Rev.* **106** (1957); J. Schwinger, *Ann. Phys.* **1** (1957).
17. T. D. Lee and C. N. Yang, *Phys. Rev.* **122**: 1954 (1961).
18. W. Lee, C. Fronsdal, R. Behrends, and J. Dreitlein, *Rev. Mod. Phys.* **34**: 1 (1962).
19. R. F. Sawyer, Princeton preprint (1962).

Group Symmetries with R-Invariance

R. E. MARSHAK

UNIVERSITY OF ROCHESTER
Rochester, New York

In this discussion, I shall consider (1) group symmetries (including R-symmetry), (2) electromagnetic consequences of these symmetries, and (3) R-invariance applied to the particle resonances.

1. GROUP SYMMETRIES

Attempts to bring order into the multifarious data in particle physics have led to the consideration of new symmetry principles of various kinds. We have been familiar for a long time, in all branches of physics, with invariance under the proper orthochronous Lorentz group and under the discrete space–time transformations such as parity (P) and time-reversal (T), which lead to multiplicative constants of motion. In particle physics, internal symmetries also play a very important role and, recently, a great deal of effort has gone into trying to enlarge the internal symmetry groups. The well-established conservation laws in particle physics thus far seem to be associated with continuous gauge transformations (which lead to additive constants of motion) or with rotations in some abstract space (e.g., isospin). The hybrid operation, charge conjugation (C), is a discrete operation connected to the discrete operations of the Lorentz group P and T (through the *CPT* theorem) and to the isospin spin group (through its inversion of the sign of I_3).* Some attention has been paid to a discrete purely internal symmetry operation which reflects the hypercharge (Y) and the electric charge (Q) and is known as R-conjugation.[1,2] In view of the promise

---

* Because of the relation between Q, Y, and I_3 imposed by the Gell-Mann–Nishijima equation $Q = I_3 + (Y/2)$.

of the SU_3 group, I should like to report on some results obtained by Dr. Okubo and myself[3] by combining SU_3 symmetry with R-invariance in several possible formulations.

We shall first consider the present situation with regard to the SU_3 group. The physical motivation for enlarging the internal group symmetries stems from the three conservation laws obeyed by strong interactions, that is, (a) hypercharge (Y), (b) I-spin (I), and (c) baryon number (B). Laws (a) and (c) can be considered as consequences of gauge transformations G_Y, G_B, while (b) may be thought of as invariance under rotations in three-dimensional space (R_3), which may be represented by the group operation SU_2. Thus, we require the strong interactions to be invariant under the product of these transformations, that is, $G_B \otimes R_3 \otimes G_Y$ (represented by the group operation $U_1^{(B)} \otimes SU_2 \otimes U_1^{(Y)}$). We must replace $SU_2 \otimes U_1^{(Y)}$ by U_2,* and thus we require invariance under $U_1^{(B)} \otimes U_2$. This is a subgroup of U_3 but strictly speaking not of SU_3.

We shall be chiefly considering the SU_3 (octet) model, but mention briefly the Sakata and global symmetry models. The Sakata model fits more properly within the framework of U_3 than SU_3. This is because we may specify B, I, Y for the Sakata model, while for the SU_3 (octet) model we can only specify I and Y.

Global symmetry is based on rotations in four-dimensional space (R_4). We have the four doublets given by

$$N_1 = \begin{pmatrix} p \\ n \end{pmatrix} \qquad N_2 = \begin{pmatrix} \Xi^0 \\ \Xi^- \end{pmatrix} \qquad N_3 = \begin{pmatrix} \Sigma^+ \\ Y^0 \end{pmatrix} \qquad N_4 = \begin{pmatrix} Z^0 \\ \Sigma^- \end{pmatrix}$$

where

$$Y^0 = \frac{\Lambda - \Sigma^0}{\sqrt{2}} \qquad \text{and} \qquad Z^0 = \frac{\Lambda + \Sigma^0}{\sqrt{2}}$$

The operation which connects

$$(N_1, N_2) \quad \text{to} \quad (N_3, N_4) \quad \text{is} \quad R_3^I$$

while (N_1, N_3) and (N_2, N_4) are connected by R_3^Y. $R_3^I \otimes R_3^Y$ is equivalent to R_4, which is a subgroup of U_4 but not U_3.

We will follow the notation of Okubo[4] and use f_1, f_2, f_3 as the parameters which characterize the different representations of U_3, the dimension of which is defined by

$$d = \tfrac{1}{2}(f_1 - f_2 + 1)(f_1 - f_3 + 2)(f_2 - f_3 + 1)$$

---

* Because of the relation between I and Y [i.e., $(-1)^{2I+Y} = 1$] imposed by the Gell-Mann–Nishijima relation.

Further, we have

$$B = f_1 + f_2 + f_3$$
$$I = \tfrac{1}{2}(f_1' - f_2')$$

and

$$Y = (f_1' + f_2') - (f_1 + f_2 + f_3)$$

where f_1 and f_2 can take on all possible values given by

$$f_1 \geqslant f_1' \geqslant f_2 \geqslant f_2' \geqslant f_3$$

Representation	Dimension	I	Y
$U_3\,(0, 0, 0)$	1	0	0
$U_3\,(1, 0, 0)$	3	$\tfrac{1}{2}, 0$	$1, 0$
$U_3\,(1, 0, -1)$	8	$\tfrac{1}{2}, \tfrac{1}{2}, 1, 0$	$1, -1, 0, 0$
$U_3\,(2, 0, -1)$	15	$\tfrac{1}{2}, \tfrac{1}{2}, \tfrac{3}{2}$	$-1, 1, 1$
		$1, 1, 0$	$2, 0, 0$
$U_3\,(1, 1, -1)$	6	$\tfrac{1}{2}, 0, 1$	$1, 2, 0$
$U_3\,(2, 0, -2)$	27	$0, \tfrac{1}{2}, \tfrac{1}{2}, \tfrac{3}{2}, \tfrac{3}{2}$	$0, 1, -1, 1, -1$
		$1, 1, 1, 2$	$0, 2, -2, 0$
$U_3\,(2, -1, -1)$	10	$\tfrac{1}{2}, \tfrac{3}{2}, 1, 0$	$-1, 1, 0, -2$
$U_3\,(1, 1, -2)$	$\overline{10}$	$\tfrac{1}{2}, \tfrac{3}{2}, 1, 0$	$1, -1, 0, 2$

In the Sakata model, the three primitive objects are taken to be (p, n, Λ) represented by $U_3\,(1, 0, 0)$ and the mesons belong to the representation $U_3\,(1, 0, -1)$, so that the higher baryon multiplets can be assigned to one of the decompositions of

$$U_3\,(1, 0, 0) \otimes U_3\,(1, 0, -1)$$
$$= U_3\,(2, 0, -1) \oplus U_3\,(1, 1, -1) \oplus U_3\,(1, 0, 0) = 15 \oplus 6 \oplus 3$$

For the octet model, the representation $U_3\,(1, 0, -1)$ is taken to contain the eight known baryons N, Λ, Σ, and Ξ. Higher baryon multiplets and mesons are formed from decompositions of the product representation

$$U_3(1, 0, -1) \otimes U_3(1, 0, -1)$$
$$= 2U_3(1, 0, -1) \oplus U_3(0, 0, 1) \oplus U_3(2, 0, -2)$$
$$\oplus U_3(2, -1, -1) \oplus U_3(1, 1, -2)$$
$$= 8 \oplus 8 \oplus 1 \oplus 27 \oplus 10 \oplus \overline{10}$$

The 10 representation seems to be the promising one to accommodate N_1^*, Y_1^*, and Ξ^*. It has the unique relation between I and Y given by $I = 1 + (Y/2)$ and it consists of

State	I	Y	Mass	J	Decay
N_1^*	$\frac{3}{2}$	1	1238	$\frac{3}{2}$	$N\pi$
Y_1^*	1	0	1385	$\frac{3}{2}$	$\Lambda\pi$
$\Xi*$	$\frac{1}{2}$	-1	1532	?	$\Xi\pi$
Ω	0	-2	1679	?	Stable

where the masses of Ξ^* and Ω have been calculated using Okubo's mass relation to first order.

The basis of any representation can be written as ψ_μ, which, for the Sakata model (for baryons) is

$$\psi_\mu = \begin{pmatrix} p \\ n \\ \Lambda \end{pmatrix}$$

Bosons which form an octet in the Sakata model are represented by the traceless tensor h_ν^μ. Thus a baryon–meson interaction may be written as

$$H_1 = \overline{\psi^\mu} \gamma_5 \psi_\nu h_\mu^\nu$$

The boson–boson interaction can be written (say, for decay of vector bosons F_ν^μ into pseudoscalar mesons h_ν^μ) as

$$H_2 = F_\nu^\mu (h_\lambda^\nu \partial h_\mu^\lambda - \partial h_\lambda^\nu h_\mu^\lambda)$$

In the octet model, let us represent baryons by N_ν^μ and antibaryons by M_ν^μ, that is,

$$M_2^1 = \overline{\Sigma^+}, \quad M_1^2 = \overline{\Sigma^-}, \quad \frac{1}{\sqrt{2}}(M_1^1 - M_2^2) = \overline{\Sigma^0}, \quad -\frac{3}{\sqrt{6}} M_3^3 = \overline{\Lambda}$$

$$M_1^3 = \overline{\Xi^-}, \quad M_2^3 = \overline{\Xi^0}, \quad M_3^1 = \bar{p}, \quad M_3^2 = \bar{n}$$

and

$$N_1^2 = \Sigma^+, \quad N_2^1 = \Sigma^-, \quad \frac{1}{\sqrt{2}}(N_1^1 - N_2^2) = \Sigma^0, \quad -\frac{3}{\sqrt{6}} N_3^3 = \Lambda$$

$$N_3^1 = \Xi^-, \quad N_3^2 = \Xi^0, \quad N_1^3 = p, \quad N_2^3 = n$$

Similarly, we may write (remembering that the π triplet is the analog of the Σ triplet, etc.)

$$h_2^1 = \pi^-, \quad h_1^2 = \pi^+, \quad \frac{1}{\sqrt{2}}(h_1^1 - h_2^2) = \pi^0, \quad -\frac{3}{\sqrt{6}} h_3^3 = \eta$$

$$h_3^1 = K^-, \quad h_1^3 = K^+, \quad h_2^3 = K^0, \quad h_3^2 = \overline{K^0}$$

In the octet model, we have the possibility of two baryon–boson couplings [remembering the occurrence of U_3 $(1, 0, -1)$ twice in the product representation of $U_3\,(1, 0, -1) \otimes U_3\,(1, 0, -1)]$ which can be written as

$$H_3 = ig\,M^\mu_\nu \gamma_5\,N^\nu_\lambda h^\lambda_\mu$$
$$H_4 = ig'\,M^\mu_\nu \gamma_5\,h^\nu_\lambda N^\lambda_\mu$$

Further, in the octet model, the components of the "10" representation have the tensor notation given by

$$F^{\mu\nu}_{\alpha\beta}$$

which is symmetric in μ, ν and antisymmetric in α, β. The "$\overline{10}$" representation $F^{\mu\nu}_{\alpha\beta}$ is antisymmetric in μ, ν and symmetric in α, β. Finally, the decay of a baryon belonging to "10" into a baryon of "8" and a meson of "8" can be represented by

$$S = F^{\mu\nu}_{\alpha\beta}\,M^\alpha_\mu\,h^\beta_\nu$$

This is unique, unlike the case when we have the decay of a baryon of "8" into a baryon of "8" and a meson of "8" for which we have two linearly independent interactions

$$S_1 = M^\mu_\nu h^\nu_\lambda N^\lambda_\mu$$
$$S_2 = M^\mu_\nu h^\lambda_\mu N^\nu_\lambda$$

Thus, the decay probability of a baryon in "10" depends on only one matrix element, whereas the decay of a baryon in "8" depends on two matrix elements.

We are now ready to introduce the operation of *R*–conjugation. *R*–conjugation has been defined for the baryon and pseudoscalar meson octets and the photon as follows:

$$R:\ p \leftrightarrow \Xi^-,\quad n \leftrightarrow -\Xi^0,\quad \Lambda \leftrightarrow \Lambda,\quad \Sigma^+ \leftrightarrow \Sigma^-,\quad \Sigma^0 \leftrightarrow \Sigma^0;$$
$$K^+ \leftrightarrow K^-,\quad K^0 \leftrightarrow \overline{K^0},\quad \eta \leftrightarrow \eta,\quad \pi^+ \leftrightarrow \pi^-,\quad \pi^0 \leftrightarrow \pi^0;\quad \gamma \leftrightarrow -\gamma \tag{1}$$

It is evident that for each particle, the operation of *R*–conjugation changes the hypercharge Y into $-Y$, the electric charge Q into $-Q$, and *a fortiori* the third component of the isospin I_3 into $-I_3$ (because of the Gell-Mann–Nishijima relation). The choice of signs in (1) is such that the *D*-type interaction [i.e., $(H_3 + H_4)$ above] between the baryon and meson octets is invariant under *R*. There is, of course, no need to stick with the signs in (1); one can alter the signs so that the operation of *R*-conjugation leaves the *F*-type interaction [i.e., $(H_3 - H_4)$ above] invariant, namely:

$$R': p \leftrightarrow \Xi^-, \quad n \leftrightarrow \Xi^0, \quad \Lambda \leftrightarrow -\Lambda, \quad \Sigma^+ \leftrightarrow \Sigma^-, \quad \Sigma^0 \leftrightarrow -\Sigma^0;$$
$$K^+ \leftrightarrow K^-, \quad K^0 \leftrightarrow -\overline{K^0}, \quad \eta \leftrightarrow -\eta, \quad \pi^+ \leftrightarrow \pi^-, \quad \pi^0 \leftrightarrow -\pi^0;$$
$$\gamma \leftrightarrow -\gamma \tag{2}$$

It is possible to write many other definitions of R-conjugation; however, (1) and (2) have the virtue that they are both consistent with unitary symmetry and that they act in opposite ways on the D and F interactions, i.e., $RD = D$, $RF = -F$; $R'D = -D$, $R'F = F$. Indeed, except for the signs, both R and R' are simply operations which interchange the covariant and contravariant indices of the tensors involved in the unitary symmetry model. This has the consequence that, e.g., the "8" representation stays "8", whereas "10" becomes "$\overline{10}$" under the operation of R-conjugation. The combination of either R or R' invariance with the SU_3 group has rather stringent consequences for strong and electromagnetic interactions and we wish to mention some of them. It is implicit in this discussion that the largest mass difference (that between N and Ξ) is neglected and that all predictions will be in error by terms of the order

$$\Delta = \frac{m_\Xi - m_N}{m_\Xi + m_N} \simeq 0.17$$

2. ELECTROMAGNETIC EFFECTS

Let us consider the electromagnetic effects first. The electromagnetic current in the Sakata model is given by

$$j_\mu = \bar{p}\gamma_\mu p$$

This obviously has the unitary transformation property of T^1_1, where

$$T^\mu_\nu = \overline{\psi^\mu}\psi_\nu$$

Define

$$\langle S^\mu_\nu \rangle = a\delta^\mu_\nu T^\lambda_\lambda + bT^\mu_\nu$$

If we require the trace to vanish, we have the condition

$$3a + b = 0$$

and hence:

$$\langle S^1_1 \rangle = a(\bar{p}p + \bar{n}n + \bar{\Lambda}\Lambda) - 3a\bar{p}p = a(\bar{n}n + \bar{\Lambda}\Lambda - 2\bar{p}p)$$

Since all the electromagnetic form factors are proportional to the current, the expression for $\langle S^1_1 \rangle$ leads immediately to the following relations for the magnetic moments, say,

$$\mu(n) = \mu(\Lambda) \qquad \mu(p) = -\frac{\mu(n)}{2}$$

If we do not impose the tracelessness condition, which would be more in the spirit of the Sakata model (U_3 group), we would only have the relation $\mu(n) = \mu(\Lambda)$.

In the SU_3 (octet) model, we can write for the electromagnetic current:

$$j_\mu = \bar{p}\gamma_\mu p + \overline{\Sigma^+}\gamma_\mu\Sigma^+ - \overline{\Sigma^-}\gamma_\mu\Sigma^- - \overline{\Xi^-}\gamma_\mu\Xi^-$$

Since we are interested only in the tensorial property in unitary space of j_μ, we may forget the γ_μ for the present and write

$$j_\mu = M_3^1 N_1^3 + M_2^1 N_1^2 - M_1^2 N_2^1 - M_1^3 N_3^1$$

Define

$$S_\nu^\mu = M_\lambda^\mu N_\nu^\lambda - M_\mu^\lambda N_\lambda^\mu$$

As far as electromagnetic properties are concerned, it is easy to see that we are again interested only in S_1^1 and that whatever relations we obtain will hold for both the charge and magnetic form factors of the baryons. For spin-0 bosons, there are no magnetic form factors but only charge form factors.

Let us write the general expectation value of S_ν^μ

$$\langle S_\nu^\mu \rangle = a\delta_\nu^\mu(M_\beta^\alpha N_\alpha^\beta) + bM_\lambda^\mu N_\nu^\lambda + cM_\nu^\lambda N_\lambda^\mu$$

If we impose the condition that the trace of this tensor is zero, we get the relation

$$3a + b + c = 0$$

Substituting, we obtain for $\langle S_1^1 \rangle$

$$\langle S_1^1 \rangle = aM_\beta^\alpha N_\alpha^\beta + bM_\lambda^1 N_1^\lambda - (3a + b)M_1^\lambda N_\lambda^1$$

Hence, the octet model yields the following relations for the magnetic form factors of the baryon octet :

$$\mu(\Sigma^+) = \mu(p)$$
$$\mu(\Lambda) = \tfrac{1}{2}\mu(n)$$
$$\mu(\Xi^0) = \mu(n)$$
$$\mu(\Xi^-) = \mu(\Sigma^-) = -[\mu(p) + \mu(n)]$$
$$\mu(\Sigma^0) = -\tfrac{1}{2}\mu(n)$$

with similar relations for the charge form factors. The implied relation $\mu(\Sigma^0) = \tfrac{1}{2}[\mu(\Sigma^+) + \mu(\Sigma^-)]$ follows simply from just charge independence. By applying the same arguments to the pseudoscalar meson octet, we find

$$F(K^0) = F(\overline{K^0})$$

where F is the charge form factor of the mesons. But we know that

$$\langle K^0 | j_\mu | K^0 \rangle = -\langle \overline{K^0} | j_\mu | \overline{K^0} \rangle$$

by charge conjugation. Thus, $F(K^0) = -F(\overline{K^0})$, which, when combined with unitary symmetry, yields $F(K^0) = 0$. Also, we have

$$F(K^+) = F(\pi^+)$$

This is quite natural, since the difference in mass between π and K is neglected to this approximation in the unitary symmetry model. It should also be pointed out that for meson octets, the antiparticles are included in the same representation—in contrast to baryon octets.

To first order in the electromagnetic interaction and to all orders in unitary symmetry, one can deduce the relation[5]

$$m_{\Xi^-} - m_{\Xi^0} = m_{\Sigma^-} - m_{\Sigma^+} + m_p - m_n$$

This equation predicts (using the latest values of the masses) 6.4 MeV for $(m_{\Xi^-} - m_{\Xi^0})$, whereas the latest experimental value is 6.5 ± 1.0 MeV. This good agreement looks like a strong argument in support of the octet model for the baryons. Unfortunately, the same model does not work as well for the meson octet. We have noted earlier that

$$F(K^+) = F(\pi^+) \qquad \text{and} \qquad F(K^0) = F(\pi^0) = 0$$

This implies that m_{K^+} is greater than m_{K^0}, since m_{π^+} is greater than m_{π^0}, which is contrary to experiment. It will be recalled that the mass formula relating the masses of π, K, and η (within the meson octet) does not work quite as well as the formula for N, Λ, Σ, and Ξ (within the baryon octet). The poorer result for the meson octet compared to that for the baryon octet may be a reflection of the fact that there are larger deviations from unitary symmetry (due to the larger mass differences within the octet) for the mesons than for the baryons.

We now impose the additional condition of R or R' invariance. We have already noted that, in the octet model, the operation of R (or R') conjugation, apart from any sign changes, merely interchanges the covariant and contravariant indices; thus

$$RN^\mu_\nu = N^\nu_\mu \qquad\qquad Rh^\mu_\nu = h^\nu_\mu$$

R symmetry is not contained in unitary symmetry and therefore may be independently defined; but if we impose both unitary symmetry and R invariance simultaneously, the consequences are severe. Thus, consider the electromagnetic current

$$\langle j_\mu \rangle = \langle S_1^1 \rangle = M_\lambda^1 N_1^\lambda - M_1^\lambda N_\lambda^1$$

Then

$$R\langle S_1^1 \rangle = -\langle S_1^1 \rangle$$

As long as we take diagonal matrix elements of j_μ, the R' operation has the same effect as R. Hence, we can say, for example

$$\langle n | j_\mu | n \rangle = -\langle \Xi^0 | j_\mu | \Xi^0 \rangle$$

so that, for the magnetic moment, R (or R') invariance implies

$$\mu(n) = -\mu(\Xi^0)$$

When we combine this result with the unitary symmetry prediction

$$\mu(n) = \mu(\Xi^0)$$

we are forced to the conclusion $\mu(n) = 0$, which is certainly incorrect. The combination of R-invariance and unitary symmetry is therefore too potent a mixture for the electromagnetic form factors of the baryons, and one can only offer the mild qualification that Δ is not negligible and the corrections may be substantial. It should be remarked that R or R' invariance by itself (without unitary symmetry) requires

$$\langle \Lambda | j_\mu | \Lambda \rangle = -\langle \Lambda | j_\mu | \Lambda \rangle$$

and hence

$$\mu(\Lambda) = 0$$

The experimental situation here is less clear than for the neutron.

A more promising electromagnetic prediction arises in connection with the π^0 decay. The reaction $\pi^0 \rightarrow 2\gamma$ is forbidden by R' (not R) invariance; the forbiddenness of this process (in the approximation $\Delta = 0$) may not be so unfavorable, since the observed lifetime of π^0 is an order of magnitude longer than one would expect from a straightforward calculation. In contrast, the reaction $\Sigma^0 \rightarrow \Lambda + \gamma$ is forbidden by both R and R' invariance, but we cannot judge the merit of this inhibition until the lifetime for Σ^0 decay is measured.

3. PARTICLE RESONANCES

One prediction that follows from R (or R') invariance (without unitary symmetry) is that the same $(\pi\Xi)$ resonance should be found as for the $I = \frac{3}{2}, J = \frac{3}{2}$ (πN) system; there is no experimental evidence for this. R (or R') invariance also predicts that[†] the reaction $Y_1^* \longrightarrow \Sigma + \pi$ is forbidden, in contrast to the reaction $Y_1^* \longrightarrow \Lambda + \pi$; this is in accord with observation.

As an example of the application of R (or R') invariance (together with unitary symmetry) to the mesons, we consider the decay of the η-meson. The η-meson has the quantum numbers $J = 0^-$ and $G = +1$. Since $G = -1$ for the π-meson, G conservation forbids the strong decay $\eta \longrightarrow 3\pi$, so that this decay proceeds via virtual electromagnetic effects. The four modes of decay of the η are therefore of the order (α is the fine structure constant)

$$\eta \longrightarrow 3\pi \sim \alpha^2$$
$$\longrightarrow 2\pi + \gamma \sim \alpha$$
$$\longrightarrow 2\gamma \sim \alpha^2$$
$$\longrightarrow \pi^0 + 2\gamma \sim \alpha^2$$

Now, R-invariance allows all four modes of decay, whereas R'-invariance prohibits (in the approximation $\Delta = 0$) the decay modes $2\pi + \gamma$ and 2γ. Therefore, if we combine R'-invariance with unitary symmetry, the decay rates become

$$\eta \longrightarrow 3\pi \sim \alpha^2$$
$$\longrightarrow 2\pi + \gamma \sim \alpha\Delta^2$$
$$\longrightarrow 2\gamma \sim \alpha^2\Delta^2$$
$$\longrightarrow \pi^0\, 2\gamma \sim \alpha^2$$

If we insert the value for Δ and take account of the relative phase volumes, we can explain the relative frequencies of the various decay modes and, in particular, the fact that the 3π mode is comparable to the $(2\pi + \gamma)$ and 2γ modes. These results would tend to favor R' over R invariance.[‡]

[†] J. J. Sakurai[2] used this as one of the chief arguments in favor of the R-invariance of the strong interactions. The transformation properties of Y_1^* are given by the following:

$$R: Y_1^{*+} \longleftrightarrow Y_1^{*-}, \ Y_1^{*0} \longleftrightarrow Y_1^{*0}; \qquad R': Y_1^{*+} \longleftrightarrow -Y_1^{*-}, \ Y_1^{*0} \longleftrightarrow Y_1^{*0}$$

with the results mentioned in the text.

[‡] A recent paper by J. B. Bronzan and F. E. Low[6] reaches the same conclusion; this is not surprising since their A-invariance is equivalent to our R'-invariance.

Similar considerations can be given for the decays of the vector meson octet ($\rho, \omega, K^*, \overline{K}^*$) and of the baryon decuplet ($N^*, Y_1^*, \Xi^*, \Omega$) if the concept of R (or R') invariance is extended in a natural way. It is conceivable that R'-invariance plus unitary symmetry is an interesting combination for particle decays but less interesting for electromagnetic form factors and masses of resonances. Or to put in another way, the corrections due to deviations from unitary symmetry plus R'-invariance may be more important for the latter than for the former.

REFERENCES

1. G. Feinberg and R.E. Behrends, *Phys. Rev.* **115**: 745 (1959).
2. J.J. Sakurai, *Phys. Rev. Letters* **7**: 426 (1961).
3. S. Okubo and R.E. Marshak, *Nuovo Cimento* **28**: 56 (1963).
4. S. Okubo, *Progr. Theoret. Phys.* **29**: 949 (1962).
5. S. Coleman and S.L. Glashow, *Phys. Rev. Letters* **6**: 423 (1960).
6. J.B. Bronzan and F.E. Low, *Phys. Rev.* **12**: 522 (1964).

Regge Poles and Resonances*

T. K. RADHA

*MATSCIENCE
Madras, India*

---

1. INTRODUCTION

Just before this discussion of Regge poles and resonances was prepared, we received information that at Brookhaven it was found that the diffraction peaks in $\pi-p$ and $K-p$ scattering do not shrink with energy, while in the case of $p-p$ scattering the peak is much smaller than heretofore believed.[1] The logarithmic shrinking of the diffraction peak[2] was a unique prediction of the Regge poles hypothesis for high-energy scattering. Since the experimental data are not yet confirmed, we shall proceed in the usual way and try to understand the possible reasons for the diffraction peak not to shrink. Since the title of this paper refers to resonances, I shall not go into the details of the analytic properties of the functions involved.

Of course, by now we know that Regge poles are generalized bound states and resonances in complex angular momentum. So, first we shall study the potential scattering in the complex angular momentum plane and then generalize it to high-energy particle scattering also.

2. POTENTIAL SCATTERING AND POLES IN THE COMPLEX ANGULAR MOMENTUM[3]

We start from the very beginning of quantum theory with the Schrödinger equation

---

* Unlike the case of symmetries of strong interactions, the situation regarding the success or failure of the application of the theory to high-energy particle scattering still remains ambiguous.

$$\Delta \psi(\vec{r}) + E\psi(\vec{r}) = V\psi(\vec{r}) \tag{1}$$

Now, to find the solution of equation (1) we split (for central symmetric potential) the wave function into a product of functions

$$\psi = \frac{R(r, l, E)}{r} P_l^m(Z)e^{im\phi} \tag{2}$$

where the $P_l^m e^{im\varphi}$ are spherical harmonics of angular momentum l (integers). R obeys

$$\frac{1}{2m}\frac{d^2R}{dr^2} + \left[E - V(r) - \frac{l(l+1)}{2\pi r^2}\right]R = 0 \tag{3}$$

We can define the scattering amplitude $f(E, \theta)$, knowing the solution of equation (1), with the following asymptotic behavior as $r \longrightarrow \infty$:

$$\psi \sim e^{i\vec{k}\cdot\vec{r}} + f(E, \theta)\frac{e^{ikr}}{r} \tag{4}$$

where θ is the angle between $\vec{k}$ and the direction in which we take the asymptotic limit $r \longrightarrow \infty$ and $Z = \cos\theta$. Then $d\Omega |f(Z, E)|^2$ is the probability of finding the particle scattered in the solid angle $d\Omega$ with outgoing momentum $\vec{k}$. Then we expand

$$f(Z, E) = \sum_{l=0}^{\infty} (2l + 1)f_l(E)P_l(Z) \tag{5}$$

with

$$f_l(E) = \frac{e^{2i\delta_l(E)} - 1}{2ik} = \frac{S_l(E) - 1}{2ik} \tag{6}$$

The phase shifts $\delta_l(E)$ of course depend on the potential.

Let us examine equations (2) and (5) a little more carefully. It is obvious that angular momentum has been quantized and we have taken only integer values of l. We need δ_l only when l is an integer to know the scattering amplitude. This of course is a natural consequence of the limitation on $|\cos\theta| < 1$.

Now, with the advent of the Mandelstam representation,[4] this limitation had to be given up. One can make experimental verification only when $|\cos\theta| < 1$; however, the crossing properties implied by the relativistic Mandelstam representation relate πN scattering and $N\bar{N} \longrightarrow \pi\pi$ (i.e., $N\bar{N} \longrightarrow \pi\pi$ is simply $\pi N \longrightarrow \pi N$ viewed in a region considered unphysical according to the definition $|\cos\theta| < 1$). Therefore, we have to use functions of a hyperbolic angle instead of $P_l(\cos\theta)$, requiring functions of noninteger and complex angular momenta. This concept is not totally unknown; the technique has been

used for years in the discussion of diffraction phenomena and the theory
of rainbow or propagation of waves around the earth.

The basic idea of the technique arises from a transformation, done
by Watson, of the Rayleigh–Faxen formula.[5] This too is successful
only if there exists an analytic function $f(l, E)$ of the complex variable
l which takes the value $f_l(E)$ when l is an integer. If so, we have

$$f(Z, E) = \frac{i}{2} \int_c \frac{(2l + 1)dl}{\sin\pi l} P_l(-Z) f(l, E) \qquad (7)$$

where C is defined as shown in Fig. 1.

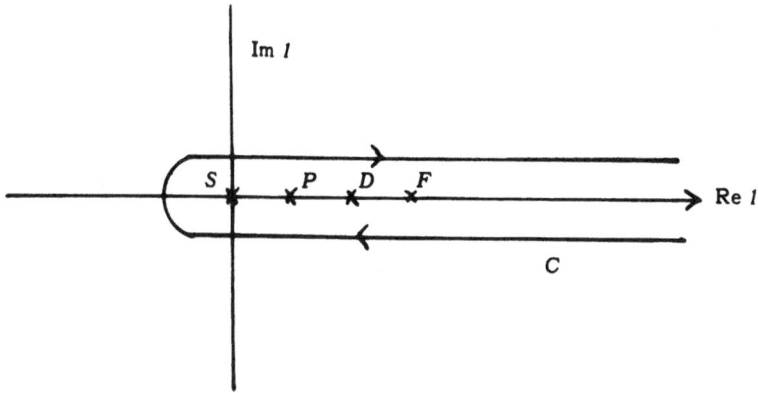

Fig. 1. The contour for the integral in equation (7).

The contour C avoids all the singularities of $f(l, E)$ and encloses
only the positive zeros of $\sin \pi l$. We can easily show that equation (7)
reduces to equation (5), since

$$\frac{1}{\sin\pi l} \sim \frac{1}{\pi(l - n)(-1)^n}$$

when $l \sim n$

$$f(l, E) = f_l(E) \text{ for integer } l$$

and by taking all the poles, we get the summation over n. The region of
convergence of equation (7) depends on P_l for complex l, and this gives
an analytic continuation outside the Lehmann ellipse within which
equation (5) was defined. Therefore, we now have to determine the
function $f(l, E)$ for general values of l and E and study its analytic
properties and asymptotic behavior for large l. Hence, we have to
study the Schrödinger equation

$$\frac{d^2R}{dr^2} + k^2R - \frac{l(l + 1)}{r^2} R - V(r)R = 0 \qquad (8)$$

Regge has shown for a superposition of Yukawa potentials[5]

$$V(r) = \int_{\mu}^{\infty} \frac{\sigma(\mu)e^{-\mu r}}{r}\, d\mu \qquad (9)$$

with

$$\int_{\mu}^{\infty} rV dr < M < \infty$$

that the scattering amplitude $f(l, E)$ is a meromorphic function of l in the half-plane Re $l > -\frac{1}{2}$, with all poles lying on the upper half of the l plane. This enables one to shift the contour C into C' (Fig. 2).

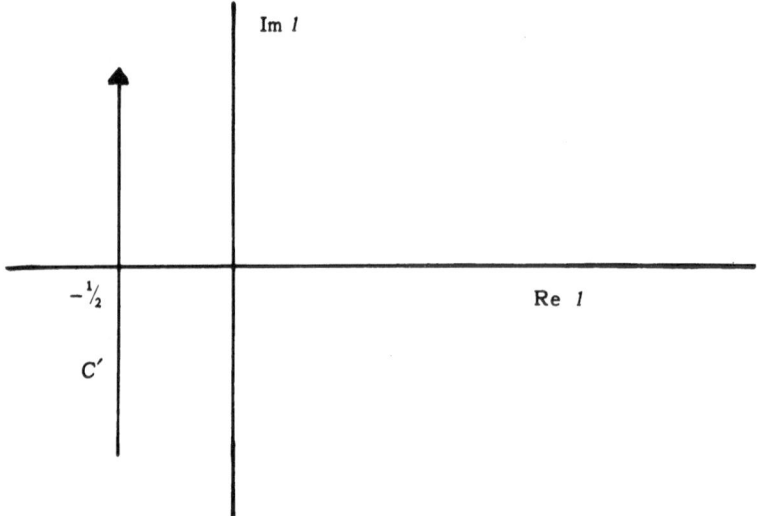

Fig. 2. The contour C' (equation 10).

We can thus finally write

$$f(Z, E) = \frac{i}{2} \int_{-1/2-i\infty}^{-1/2+i\infty} \frac{(2l+1)dl}{\sin \pi l} f(l, E) P_l(-Z)$$

$$+ \sum_{n=1}^{N} \frac{\beta_n(E)}{\sin \pi \alpha_n(E)} P_{\alpha_n(E)}(-Z) \qquad (10)$$

where the $\beta_n(E)$ are the residues of $f(\alpha_n, E)$ at the poles $l = \alpha_n(E)$ in the complex l plane. These poles which occur in the complex angular momentum plane are called the Regge poles. (The Sommerfeld–Watson representation is, strictly speaking, valid only for positive kinetic energy, but we may include bound states also by an analytic continuation in E.)[6] The first term is called the "background term,"

Regge Poles and Resonances

which vanishes as $Z \to \infty$, while the pole terms are proportional to $Z^{\alpha_n(E)}$. Now we can write the contribution to the scattering amplitude from a Regge pole at $\alpha_1 (E)$ as

$$f(Z, E) = \frac{\beta_1(E)P_{\alpha_1(E)}(-Z)}{\sin \pi \alpha_1(E)} \qquad (11)$$

We can now project out any partial wave ($l = 0$ and integer) by using

$$\tfrac{1}{2} \int_{-1}^{1} P_l(Z)P_\alpha(-Z)dZ = \frac{1}{\pi} \frac{\sin \pi \alpha}{(\alpha - l)(\alpha + l + 1)} \qquad (12)$$

and therefore

$$f_l(E) = \frac{1}{\pi} \frac{\beta_1(E)}{(\alpha - l)(\alpha + l + 1)} \qquad (13)$$

If $\alpha(E)$ for a particular E_r is close to the integer m, then we can expand it in the neighborhood of E_r as

$$\alpha(E) \approx m + \left(\frac{d \operatorname{Re} \alpha}{dE}\right)_{E = E_r} (E - E_r) + i(\operatorname{Im} \alpha)_{E = E_r} \qquad (14)$$

and therefore

$$f(l, E) = \frac{1}{\pi} \frac{\beta(E_r)/[\alpha(E_r) + l + 1]}{(m - l) + (E - E_r)(d \operatorname{Re} \alpha/dE)_{E_r}} + i \operatorname{Im} \alpha(E_r) \qquad (15)$$

which for $l = m$ has the familiar Breit–Wigner form

$$f(l, E) = \frac{\beta(E_r)/(2l + 1)(d \operatorname{Re} \alpha/dE)_{E_r}}{(E - E_r + i\Gamma/2)} \qquad (16)$$

with a width

$$\frac{\Gamma}{2} = \frac{\operatorname{Im} \alpha(E_r)}{(d \operatorname{Re} \alpha/dE)_{E_r}} \qquad (17)$$

Regge has also proved for a superposition of Yukawa potentials that Im α is positive for $E > 0$ and vanishes for $E < 0$. Thus, the Regge poles represent resonances with positive width when E is greater than 0 and bound states with $\Gamma = 0$ when E is less than 0.

3. MOVEMENT OF POLE WITH ENERGY AND POTENTIAL IN THE *l*-PLANE

We do not know what happens to the left of Re $\alpha = -\tfrac{1}{2}$, but for an attractive potential a particular pole passes through Re $\alpha = -\tfrac{1}{2}$ at some negative E and moves to the right along the real axis as E increases. When it reaches the threshold energy, the pole moves into the upper half-plane and continues its rightward movement further, but eventually swings back and reaches Re $\alpha = -\frac{1}{2}$ again. Lovelace has determined the Regge trajectories for different strengths of potentials. When $\alpha(E)$ passes through Re l = integer and Im l = 0, we have bound states, and when Re l = integer and Im l = small positive value, we have resonances (Fig. 3).

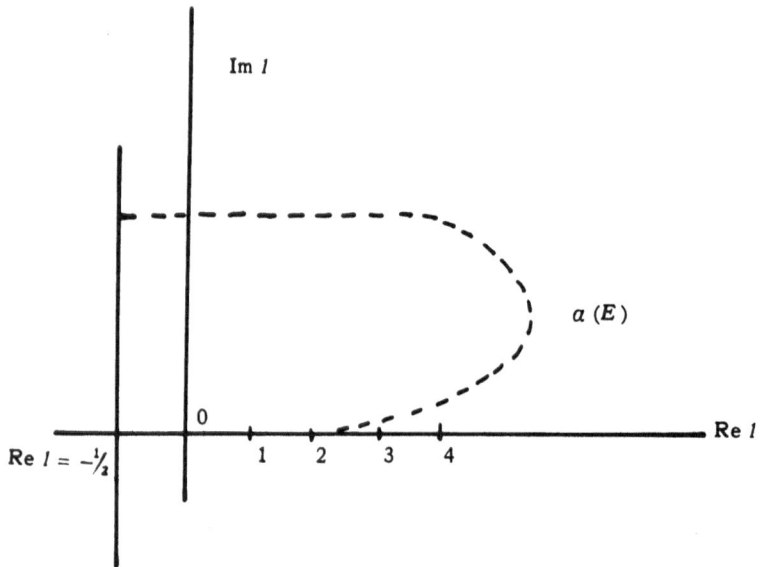

Fig. 3. A typical Regge trajectory.

We see that as the potential strength increases, the rightward movement of the pole will be extended, and we have more bound states and resonances. The same trajectory may pass through (or near) different integer values of Re l; thus we have one Regge pole giving rise to resonances and bound states in different angular momentum states (Re l = 1, 2, 3 . . .) at different energies.

So far we have not considered the exchange potential (for the space coordinates) in the Schrödinger equation. If we admit that also, there are two possibilities—symmetric (even) and antisymmetric (odd) solutions with

$$f(Z, E) = \frac{\beta(E)}{\sin \pi\alpha(E)} \frac{1}{2} [P_{\alpha(E)}(-Z) \pm P_{\alpha(E)}(Z)] \qquad (18)$$

The Regge terms corresponding to physical states of even l take the positive sign and they have positive signature[8]; likewise, terms corresponding to physical states of odd l have odd signature. Thus we have the "distance" between bound states (resonances) given by $\Delta l = 2$, if each trajectory with specific signature is considered.

APPLICATION TO THEORY OF ELEMENTARY PARTICLES

We shall extend the concept of Regge poles to relativistic two-body scattering cases like

$$p_1 + p_2 \longrightarrow p_3 + p_4$$

with

$$s = (p_1 + p_2)^2$$
$$t = (p_1 - p_3)^2 \tag{1}$$

For the equal mass case

$$\cos \theta_s = 1 - \frac{2t}{4M^2 - s} \sim t \text{ for } t \to \infty$$
$$\cos \theta_t = 1 - \frac{2s}{4M^2 - t} \sim s \text{ for } s \to \infty \tag{2}$$

The matrix element is given by

$$\langle f | S | i \rangle = \langle f | i \rangle + \frac{i}{(2\pi)^2} \frac{\delta(p_1 + p_2 - p_3 - p_4)}{(16 p_{10} p_{20} p_{30} p_{40})^{1/2}} A_{fi} \tag{3}$$

The differential cross-section is given by

$$\frac{d\sigma}{d(\cos \theta_s)} = \frac{|A|^2}{16.2\pi s}$$
$$\frac{d\cos \theta_s}{dt} = \frac{1}{s} \tag{4}$$

and by the optical theorem

$$\sigma_{\text{tot}}(s) \sim \frac{1}{s} \text{Im} A(s, 0) \tag{5}$$

Experimental facts known so far are:
1. $\sigma_{ab} = \sigma_{a\bar{b}}$ for $s \to \infty$.
2. $\sigma_{ab}^I = \sigma_{ab}^{I'}$ for $s \to \infty$.
3. All elastic scattering amplitudes show a characteristic diffraction pattern with a forward peak.
4. Width of the peak shrinks logarithmically (?).
5. At high energies, scattering amplitudes are purely imaginary.

Statements (3) and (4) lead to

$$A(s, t) \sim \beta(t)s^{\alpha(t)} = \beta(t)e^{\alpha(t)\log s} \tag{6}$$

and if $\alpha(0) = 1$ then $\sigma_t \sim$ constant. The form (6) naturally makes one wonder whether we can extrapolate the Regge results of potential scattering to relativistic particle scattering also. So we will consider the equation in the t-channel

$$A(s, t) = f(Z, E) = \sum_n \frac{\beta_n(t)P_{\alpha_n(t)}(-\cos \theta_t)}{\sin \pi\alpha_n(t)} \left\{ \frac{1 \pm e^{-i\pi\alpha_n(t)}}{2} \right\} \tag{7}$$

plus the $\int^t$ term, which vanishes as $Z \longrightarrow \infty$.

As $s \longrightarrow \infty$ (i. e., $\cos \theta_t \longrightarrow \infty$), we get

$$A(s, t) \sim \sum_n \frac{\beta_n(t)s^{\alpha_n(t)}}{\sin \pi\alpha_n(t)} \left\{ 1 \pm \frac{e^{-i\pi\alpha_n(t)}}{2} \right\} \tag{8}$$

Let us suppose that one of the Regge poles dominates so that we have (say, α_1)

$$A(s,t) \sim \frac{\beta_1(t)}{\sin \pi\alpha_1(t)} s^{\alpha_1(t)} \left\{ \frac{1 \pm e^{-i\pi\alpha_1(t)}}{2} \right\} = F(t)s^{\alpha_1(t)} \tag{9}$$

This, as we know, corresponds to the unphysical region of the t-channel. Now *to study the scattering amplitudes in the physical region we invoke crossing symmetry, which is a special feature of relativistic*

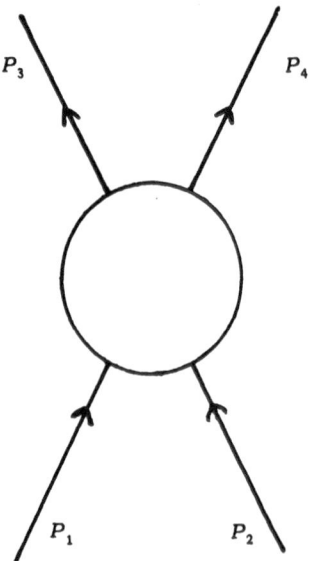

Fig. 4. Diagram for the scattering of two particles.

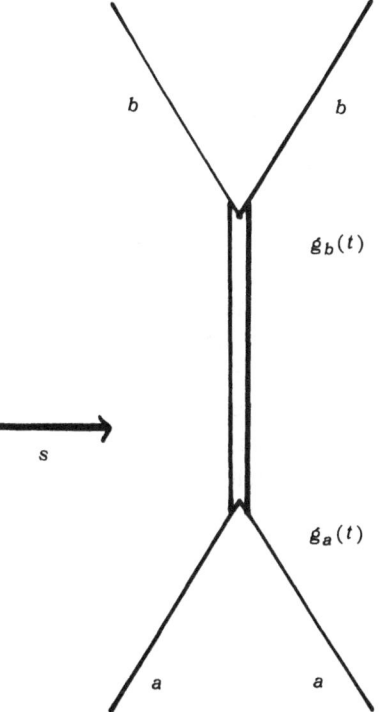

Fig. 5. Dispersion theoretic graph with pole in the t-channel.

scattering only. The same amplitude also represents scattering in the s-channel for which the region $s \to \infty$ for fixed $|t|$ is in the physical region and gives the behavior $F(t) s^{\alpha(t)}$. Further, we have near a pole in the t-channel

$$t \to \ \succ\!\!\sim\!\!\sim\!\!\sim\!\!\prec \atop \alpha(t)$$

$$A(s, t) = \frac{\beta(t_r) P_{\alpha(t_r)}(\cos \theta_t)}{\pi \operatorname{Re} \alpha'(t_r)\{t - t_r + [i \operatorname{Im} \alpha(t_r)]/[\operatorname{Re} \alpha'(t_r)]\}} \sim F(t) s^{\alpha(t)} \quad (10)$$

Let us consider the corresponding dispersion theoretic graph (Fig. 5). This gives

$$A(s, t) = \frac{g_a(t) g_b(t)}{t - m_r^2} P_l(\cos \theta_t) \sim F(t) s^l \quad \text{as } s \to \infty \quad (11)$$

where l is the angular momentum of x and $m_r^2 = t_r - (i\Gamma/2)$ where Γ is the width of the resonance

$$\Gamma = \operatorname{Im} \alpha(t_r)/\operatorname{Re} \alpha'(t_r)$$

Thus we have for asymptotic regions the behavior given by

$$\text{D.T.} \quad F(t)s^l \tag{12}$$

$$\text{R.P.} \quad F(t)s^{\alpha(t)} \tag{13}$$

The two expressions coincide only at the pole. We also know that it is the s^l behavior which persists even at large s for $l > 1$ that leads to the divergence difficulties in dispersion theory. On the other hand, Froissart[9] has proved, using Mandelstam representation, that for unitarity to be satisfied the power of s should be less than or equal to 1 for $t \leq \infty$. We at once realize that the Regge form may satisfy this criterion if $\alpha(t) < 1$ for $t \leq 0$, while for dispersion theory we need a damping. So we require that resonances and bound states with $l > 1$ are represented by Regge poles. Now it is tempting to speculate that even poles with $l = 0$ lie on Regge trajectories, in analogy to the dynamic resonances. This behavior has been conjectured by Blankenbecler and Goldberger[10] for the nucleon and for all particles by Chew and Frautschi.[11] They have tried to fit all known particles into different Regge trajectories characterized by different quantum numbers S, B, I, and parity (and signature).

Figures 6 and 7 are diagrams of Regge trajectories of baryons and baryon resonances and meson and meson resonances, respectively. The two points N and N^{***}, which can belong to the same trajectory, give a slope of approximately $1/(50\, m_\pi^2)$, which according to the Regge formula $(d/dp^2)\,(l + \frac{1}{2})^2 \approx a^2$ gives $1/(2m_\pi)$ as the radius of interaction for the bound state. Also,[13] N_{33}^* (1920 MeV) and N_1^* (1238 MeV) give a slope of $1/(50\, m_\pi^2)$, and Y_0^* (1815 MeV) and Λ (1115 MeV) give a slope of $1/(50\, m_\pi^2)$.

It may be noted that the trajectory called P or α_{vac} has no known particles on it. But one may easily identify this as the boundary beyond which no trajectory should exist, according to Froissart's theorem, if it passes through $\alpha(0) = 1$. From the diagrams, it is interesting to note that trajectories with lower quantum numbers (S, B, I) imply bigger $\alpha(0)$, and hence α_{vac} should have the simplest quantum numbers, namely, $S = B = I = 0$ of vacuum. The main features of these diagrams are: (1) The prediction of an $I = 0$, $J = 2$, $B = 0$, $S = 0$ resonance on the Pomeranchuk trajectory, for which there now seems to be experimental evidence.† (2) The "ghost" pole, which corresponds to $\alpha = 0$ at $-50\, m_\pi^2$ (-1 GeV), negative mass.

---

† This $J = 2$, $I = 0$ particle was first suggested by Lovelace.

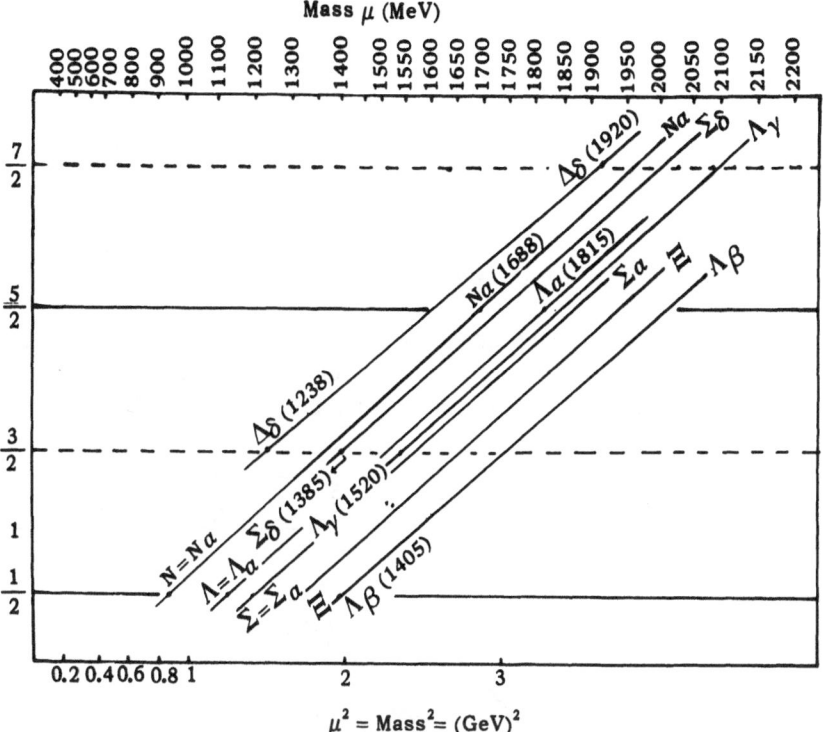

Fig. 6. Regge trajectories of baryons and baryon resonances.

We can easily see from formula (8) that if we assume the α_{vac} (P) to dominate over all other Regge (exchange) poles, then we have for large s

$$A(s, 0) \approx \lim_{t \to 0} \beta_P(t) \left[\frac{1 + e^{i\pi\alpha_P(t)}}{\sin \pi\alpha_P(t)} \right] s^{\alpha_P(t)} \approx igs^1 \qquad (14)$$

with $\alpha_P(0) = 1$ and therefore σ_{total} for the s-channel equals g equals a constant where $g = \beta_P(0)$. Thus, we find that the vacuum trajectory leads us to the Pomeranchuk theorem for $s \to \infty$. If we write

$$\alpha_P(t) = 1 + t\alpha'_P(0) \qquad (15)$$

we get from

$$\frac{d\sigma}{dt} = F(t)s^{2[\alpha_P(t)-1]}$$

$$\frac{d\sigma}{dt} \approx F(t) \exp\left[2t\alpha'_P(0) \log s\right] \qquad (16)$$

with $\alpha'(0) > 0$. This is the physical region of the s-channel (i.e.,

$t < 0$). This leads to a logarithmic shrinking of the diffraction peak with energy. However, if the Brookhaven experiments turn out to be correct, it would contradict this asymptotic behavior of the differential cross-section and may indicate the existence of cuts or immovable singularities in the angular momentum plane which may cancel the logarithmic shrinking in the case of πp and Kp diffraction scattering.

4. GELL-MANN'S FACTORIZATION PRINCIPLE

For Fig. 4 we write the amplitude in ordinary perturbation calculation as in equation (11), where we have factored the coupling constants at the two vertices. This factorizing has been established in potential

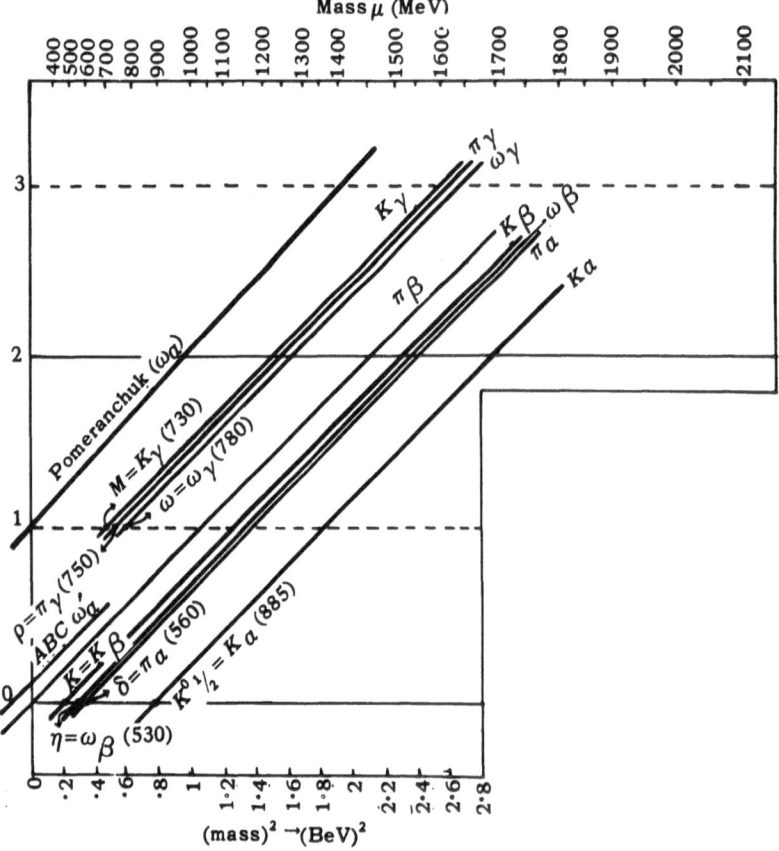

Fig. 7. Regge trajectories of mesons and meson resonances.

theory for the Regge poles by Gell-Mann[13] and for analytically continued partial wave amplitude by Gribov and Pomeranchuk.[14] According to this principle, if the same Regge pole (trajectory) P dominates all high-energy reactions ab, aa, and bb, then we have

$$\sigma_{ab}^2 = \sigma_{aa} \cdot \sigma_{bb} \qquad (17)$$

that is, we should have

$$\sigma_{\pi\pi} = \frac{\sigma_{\pi N}^2}{\sigma_{NN}} \approx 15 \text{ mb}$$

$$\sigma_{KK} = \frac{\sigma_{KN}^2}{\sigma_{NN}} \approx 8 \text{ mb, etc.} \qquad (18)$$

Thus, all polarization effects in πN scattering vanish as $s \rightarrow \infty$, and hence polarization experiments are valuable to find the interference of other poles with the α_p.

Let us consider the trajectories of other poles, i.e., resonances.[15] For this, we consider linear combination of cross-sections in order to subtract out the influence of the vacuum pole. For example, let us consider πN scattering. Then we have the A^+ and A^- amplitudes given by

$$A^+ = (\tfrac{1}{3} A^{1/2} + 2A^{3/2})$$

$$A^- = (\tfrac{1}{3} A^{1/2} - A^{3/2}) \qquad (19)$$

By optical theorem

$$\frac{1}{2} [\sigma(\pi^+ p) + \sigma(\pi^- p)] = \frac{1}{s} \text{Im } A^+ \quad (t = 0)$$

$$\frac{1}{2} [\sigma(\pi^- p) - \sigma(\pi^+ p)] = \frac{1}{s} \text{Im } A^+ \quad (t = 0) \qquad (20)$$

We also know that A^+ represents the $I = 0$, $G = +1$ channel of $\pi\pi \rightarrow N\bar{N}$ and A^- represents $I = 1$, $G = +1$ of $\pi\pi \rightarrow N\bar{N}$ (Fig. 8). Thus, we may have the vacuum and ABC pole contributing to A^+, and therefore we will have

$$\sigma(\pi^- p) + \sigma(\pi^+ p) \sim a + bE^{-[1-\alpha_{ABC}(0)]}$$

If

$$\alpha_{ABC}(0) = 0$$

we get

$$\sigma(\pi^- p) + \sigma(\pi^+ p) = a + \frac{b}{E}$$

Similarly

$$\sigma(\pi^- p) - \sigma(\pi^+ p) \propto E^{-[1-\alpha_P(0)]} \qquad (21)$$

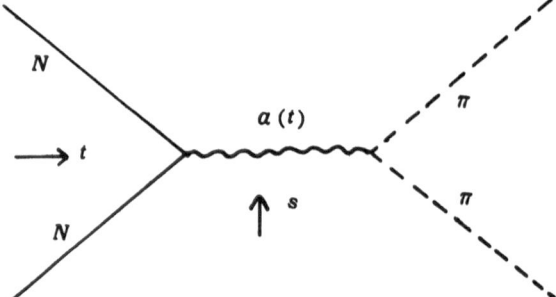

Fig. 8. Diagram representing an intermediate Regge state with spin $\alpha(t)$ in the t-channel.

Listed in Table I are the contributions of the different Regge trajectories to the various total cross sections at high energies. The π, η do not contribute at zero-momentum transfer. We can understand the signs of the different terms since the diffraction channel represents attraction, while ω and ρ (vectors) represent repulsion for identical particles and attraction between particle and antiparticle. Remaining signs follow from isotopic spin considerations.

The ω and ρ Trajectory

We can deduce from the table that

$$\frac{1}{2}\left[\sigma(\bar{p}p) - \sigma(pp)\right] = \pi\epsilon_\omega g^2_{pp\omega}\left(\frac{s}{s_0}\right)^{\alpha_\omega(0)-1} + \pi\epsilon_\rho g^2_{pp\rho}\left(\frac{s}{s_0}\right)^{\alpha_\rho(0)-1} \quad (22)$$

$$\frac{1}{2}\left[\sigma(np) - \sigma(pp)\right] = \pi\epsilon_\rho g^2_{pp\rho}\left(\frac{s}{s_0}\right)^{\alpha_\rho(0)-1} \quad (23)$$

In the 10 GeV region $\frac{1}{2}[\sigma(p\bar{p}) - \sigma(pp)] = 20$ mb, while $\frac{1}{2}[\sigma(np) - \sigma(pp)] = 2$ mb. Therefore, we may neglect the contribution of the ρ trajectory in equation (22), as can be deduced from equation (23), and concentrate on the ω in interpreting the $\sigma(pp) - \sigma(pp)$ difference. Thus, from experiments, we find $\alpha_\omega(0) \approx 0.4$, which agrees with the value given by the Chew and Frautschi diagram. Neglecting the ρ and ABC contribution, we have

$$\sigma(pp) = \sigma(\infty) - \pi\epsilon_\omega g^2_{pp\omega}\left(\frac{s}{s_0}\right)^{\alpha_\omega(0)-1}$$

$$\sigma(p\bar{p}) = \sigma(\infty) + \pi\epsilon_\omega g^2_{pp\omega}\left(\frac{s}{s_0}\right)^{\alpha_\omega(0)-1} \quad (24)$$

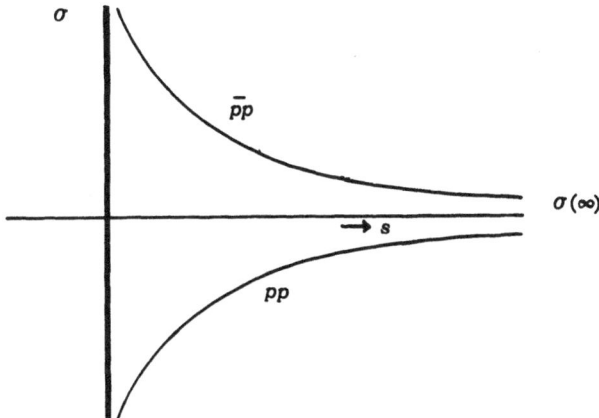

Fig. 9. Total cross sections for p-p and p-$\bar{p}$ scattering.

And we expect the behavior shown in Fig. 9 for $\sigma(pp)$ and $\sigma(p\bar{p})$. Then why is σ_{pp} approaching a constant value (40 mb) beyond 10 GeV, while it is not constant in this region? Perhaps $\sigma(pp)$ constant means that there is a third trajectory P' which just cancels the imaginary part of the trajectory in the pp amplitude and adds to it for $p\bar{p}$ scattering. And this vacuum trajectory with

$$\alpha_{P'}(0) = \alpha_{\omega}(0) = 0.4$$

would satisfy the above requirements, that is

$$\sigma(pp) = \sigma(P) - \sigma(\omega) + \sigma(P') \tag{25}$$

$$\sigma(p\bar{p}) = \sigma(P) + \sigma(\omega) + \sigma(P') \tag{26}$$

If the ABC is responsible for this, then its trajectory cannot be simple like the others, but may have a strange twist. Similar results for K^+p and K^-p total cross sections hold: that is, $\sigma(K^+p) \approx 18$ mb above 5 GeV, while $\sigma(K^-p)$ is not a constant. Thus, the requirement for a P' seems more reasonable. Also this requires that the ratio of the couplings of P' and ω with K mesons and nucleons be the same.

In the case of π^-p, π^+p scattering, ω cannot contribute since $G = -1$ for ω. Thus

$$\sigma(\pi^+p) + \sigma(\pi^-p) = \sigma(P) + \sigma(ABC) + P' \text{ contribution} \tag{27}$$

$$\sigma(\pi^-p) - \sigma(\pi^+p) = \pi\epsilon_{\rho}g_{\pi\pi\rho}g_{NN\rho}\left(\frac{s}{s_0}\right)^{\alpha_{\rho}(0)-1} \tag{28}$$

Table I. Regge Pole Contribution to Various Cross Sections

Cross section	Regge poles	Expected high-energy behavior
$\sigma(\pi^- p) + \sigma(\pi^+ p)$	P, ABC	$2\left[\sigma(\infty) + \pi\varepsilon_{ABC}g_{\pi\pi ABC}g_{ppABC} \times \left(\frac{s}{s_0}\right)^{\alpha_{ABC}(0)-1}\right]$
$\sigma(\pi^- p) + \sigma(\pi^+ p)$	ρ	$2\pi\varepsilon_\rho g_{\pi\pi\rho}g_{ppp}\left(\frac{s}{s_0}\right)^{\alpha_\rho(0)-1}$
$\sigma(pp) + \sigma(\bar{p}p)$	P, ABC	$2\left[\sigma(\infty) - \pi\varepsilon_{ABC}g^2_{ppABC}\left(\frac{s}{s_0}\right)^{\alpha_{ABC}(0)-1}\right]$
$\sigma(pp) - \sigma(\bar{p}p)$	ω, ρ	$2\left[\pi\varepsilon_\omega g^2_{\bar{p}p\omega}\left(\frac{s}{s_0}\right)^{\alpha_\omega(0)-1} + \pi\varepsilon_\rho g^2_{\bar{p}pp}\left(\frac{s}{s_0}\right)^{\alpha_\rho(0)-1}\right]$
$\sigma(pp) - \sigma(np)$	ρ	$2\pi\varepsilon_\rho g^2_{\bar{p}pp}\left(\frac{s}{s_0}\right)^{\alpha_\rho(0)-1}$
$\sigma(pp) + \sigma(np)$	P, ABC, ω	$2\left[\sigma(\infty) + \pi\varepsilon_{ABC}g^2_{ppABC}\left(\frac{s}{s_0}\right)^{\alpha_{ABC}(0)-1} - \pi\varepsilon_\omega g^2_{pp\omega}\left(\frac{s}{s_0}\right)^{\alpha_\omega(0)-1}\right]$
$\sigma(K^- p) - \sigma(K^+ p)$	ω, ρ	$2\left[\pi\varepsilon_\omega g_{KK\omega}g_{pp\omega}\left(\frac{s}{s_0}\right)^{\alpha_\omega(0)-1} - \pi\varepsilon_\rho g_{KK\rho}g_{ppp}\left(\frac{s}{s_0}\right)^{\alpha_\rho(0)-1}\right]$

and they approach the Pomeranchuk limit as the contribution from P' vanishes as $s \rightarrow \infty$. Again, one gets $\alpha_{P'}(0) \approx 0.4$. It was Igi[16] who first suggested the P' trajectory while treating non-charge-exchange πN scattering when he assumed the amplitude could be written as sum of two terms—one due to P and other giving an amplitude which decreases with s and satisfies an unsubtracted dispersion relation. But this gives a different scattering length from the observed value, and hence we need either another subtraction due to P' which varies as $s^{\alpha(0)}$ where $\alpha(0) \sim 0.5$, or admitting cuts in the angular momentum plane.

The ρ Trajectory

To evaluate $\alpha\rho$ we have to study

$$\sigma(pn) - \sigma(pp) \quad \text{or} \quad \sigma(\pi^- p) - \sigma(\pi^+ p)$$

as given in Table I.

The ρ coupling to N is much weaker than the ω. The sign of $\sigma(pn) - \sigma(pp)$ difference is important and also the magnitude. $\sigma(pn) - \sigma(pp)$ has also a contribution from π exchange which may be smaller than ρ. But $\sigma(\pi^- p) - \sigma(\pi^+ p)$ is due to ρ alone. This analysis leads to $\alpha\rho(0) = 0.3$. However, $\Delta\sigma \approx 1.5$ mb from 10 to 20 GeV. This may be a violation of the Pomeranchuk limit that $\Delta\sigma \rightarrow 0$, and in this case we may get $\alpha\rho(0) = \alpha(t_r) = 1$, which indicates an unreggeized behavior. Little is understood about the ρ.

Hahn[17] has shown that

$$\sigma(pn \rightarrow np) \sim 100 \text{ mb}$$

and

$$\frac{d\sigma(np \rightarrow pn)}{d\Omega}$$

is isotropic up to

$$|t| \sim 2 \text{ GeV}^2$$

at

$$E_{\text{lab}} = 25 \text{ GeV}$$

and this leads to a $\sigma(np) - \sigma(pp)$, which is an order of magnitude larger than that predicted by ρ alone. One wonders whether the π has a significant part to play in this. As a matter of fact, an elastic unreggeized π exchange graph of Ferrari and Selleri gives the observed values correctly. Let us list a number of reactions in which the Regge pole hypothesis can be tested:

$$N \qquad\qquad \pi + N \rightarrow N + \pi$$
$$\gamma + N \rightarrow N + \pi$$
$$\pi + N \rightarrow N + \omega \qquad \text{etc.}$$

$$Y = \Lambda, \Sigma \qquad \pi + N \rightarrow Y + K$$
$$K + N \rightarrow N + K$$
$$Y + N \rightarrow Y + K \qquad \text{etc.}$$

$$\pi \qquad\qquad N + N \rightarrow N + N$$
$$Y + N \rightarrow \pi + N \qquad\qquad (29)$$
$$\pi + N \rightarrow \rho + N$$
$$K + N \rightarrow K^+ + N$$
$$N + N \rightarrow N + N^*$$

$$K \qquad\qquad \pi + N \rightarrow K^* + Y$$
$$\gamma + N \rightarrow K + Y$$

$$\omega \qquad\qquad N + N \rightarrow N + N$$
$$\pi + N \rightarrow \rho + N$$

Note that we will have equality of the differential cross section for $\pi^- p \rightarrow K^0 \Lambda^0$ and $K^0 + p \rightarrow \pi^+ + \Lambda$ when the same pole dominates. Finally, the present conclusions on the basis of no Regge cuts as summarized by Drell at the CERN conference are

1. $\alpha_P(0) = 1, \quad \alpha_{P'}(0) \approx 0.4, \quad \alpha_\omega(0) \approx 0.4.$
2. ρ is weakly coupled and is unknown.
3. The fermion trajectories are unknown.
4. π is yet to be explored.

The Hahn experiment shows un-Reggeized π behavior. But Frautschi *et al.* seem to deduce a Reggeized behavior for π also from the analysis of $N + N \rightarrow N + N^*$.

5. THEORETICAL DEVELOPMENTS

1. Using elastic unitarity, Mandelstam[18] and Gribov[19] have shown that the only singularities (energy dependent) of the analytically continued partial wave amplitudes $f_l(s)$ in the angular momentum plane are just simple poles. However, this does not rule out the possibility of poles at fixed (energy independent) values of the angular momentum.
2. Eden[20] has also come to the same results by replacing the as-

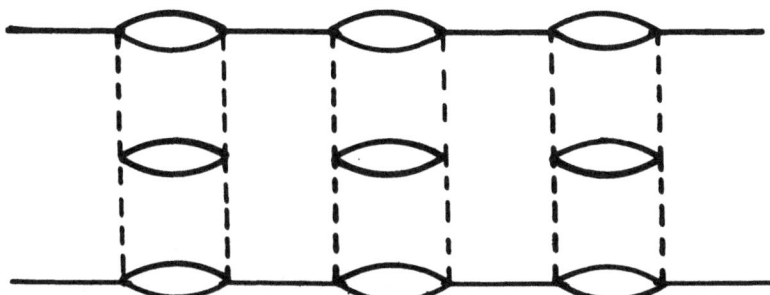

Fig. 10. Graphs contributing to the absorptive amplitudes.

sumption of elastic unitarity by the condition of nonvanishing of the imaginary part of the trajectory $l(s)$ in the physical region.

3. Lee and Sawyer[21] have shown that the Bethe–Salpeter scattering amplitude in the ladder approximation has at least one Regge pole in the region $\mathrm{Re}\, l > -\frac{3}{2}$, and further that the scattering amplitude has only poles in this region.

4. Using model field theory, Amati et al.[22] also arrived at the Regge behavior for the scattering amplitude, including the possibility of a "ghost" with $\alpha(t) = 0$ for some $t < 0$. To avoid the "ghost," Gell-Mann has suggested that the coupling of the trajectory to any system is proportional to $\alpha(t)$, so that there is no residue at the "ghost." This is quite welcome, because we need $\alpha(t = -\infty) < 0$ to avoid subtractions in dispersion relations, and we must have $\alpha(t) = 0$ at some t if we want $\alpha(0) = +1$, which we require for constant total cross-sections. This behavior also exists in the case of Regge poles for nuclei[23] with ground state $J > \frac{3}{2}$, say, 2. Then, before we go to negative values of α, we will have to pass the $\alpha = 0$ state, but this cannot exist, and so we again assume that the coupling goes to 0.

Amati et al. consider that the ghost is due to the fact that the scattering amplitude is not unitary in the s-channel. So they added graphs of the type shown in Fig. 10 to insure unitarity in the s-channel also. But this leads to the following behavior:

$$A_i(s, t) \sim f(t)s^{\alpha(t)} + \frac{g(t)}{\log s} s^{\alpha_M(t)} \qquad (30)$$

The second term indicates the existence of a cut in the angular momentum plane from $\alpha_M(t) \geq \alpha(t) + \alpha(0) - 1$. Further, one finds that $\alpha_M(0) = \alpha(0) = 1$, and then the second term becomes comparable to the first for

$$|t| \sim \frac{\log (\log s)}{\log s} \tag{31}$$

because asymptotically $\alpha_M(t) \approx 2\alpha(t/4) - 1$.

This model is applicable only between 0 and $|t|$ where

$$|t| = \frac{\log (\log s)}{\log s} \quad \text{for } s \to \infty \tag{32}$$

and we may escape the "ghost." But then the approach to asymptotic σ is only logarithmic in s. The way out suggested by many people is that complete restoration of unitarity in the t-channel (outside the strip region) may remove these cuts. Gell-Mann and Udgoankar[23] have suggested that cuts occur *only* when the particles participating have anomalous thresholds like nuclei. The need for the cut was recognized by them on the basis of a factoring principle on the coupling of Regge poles.

This principle is common to all field theoretic models in which there is just one dominant channel for the interaction.

5. Bardacki has given a proof for the existence of Regge poles in relativistic S–matrix theory for Re $l > 1$.

Topics of interest to be explored can be now summarized:

1. Scattering of particles by resonances and the definition of Clebsch–Gordan coefficients to complex values of the indices (l).
2. Identification of the interpolation function $f(l, k)$ with partial wave amplitude $f_l(k)$ for all l. Coincidence up to $l = 0$ seems to be all right, but for $l = 0$ it is very difficult to prove.
3. Extension of the multiperipheral model to three-body problems.
4. The question of the existence of Regge cuts. This may be the reason for the constancy of the diffraction peak width for π–p and K–p scattering.
5. Study of Regge poles for complex potentials.
6. Possible connections between symmetries and Regge poles in high-energy scattering.

REFERENCES

1. A.P. Balachandran, Private communication.
2. Diddens *et al.*, Report to CERN Conference (1962).
3. T. Regge, Lectures delivered at Summer School held at Trieste (1962).
4. S. Mandelstam, *Phys. Rev.* **112**: 1344 (1958).

5. T. Regge, *Nuovo Cimento* **14**: 951 (1959); **18**: 947 (1960). Also, A. Bottino, A.M. Longoni, and T. Regge, *Nuovo Cimento* **23**: 954 (1962).

6. G.F. Chew, S. Mandelstam, and S.C. Frautschi, *Phys. Rev.* **126**: 1202 (1962).

7. C. Lovelace and D. Masson, University of London, Preprint and report to CERN Conference (1962).

8. M. Gell-Mann, S.C. Frautschi, and F. Zachariasen, *Phys. Rev.* **126**: 2204 (1962).

9. M. Froissart, *Phys. Rev.* **123**: 1053 (1961).

10. R. Blankenbecler and M.L. Goldberger, *Phys. Rev.* **126**: 766 (1962).

11. G.F. Chew and S.C. Frautschi, *Phys. Rev. Letters* **7**: 394 (1961); **8**: 41 (1962).

12. A.H. Rosenfeld, UCRL-10492.

13. M. Gell-Mann, *Phys. Rev. Letters* **8**: 263 (1962).

14. V.N. Gribov and I. Pomeranchuk, *Phys. Rev. Letters* **8**: 349 and 412 (1962).

15. B.M. Udgoankar, *Phys. Rev. Letters* **8**: 142 (1962).

16. K. Igi, Preprint, Berkeley.

17. B. Hahn *et al.*, Report to CERN Conference (1962).

18. S. Mandelstam, Report to CERN Conference (1962).

19. V.N. Gribov, Report to CERN Conference (1962).

20. R. Eden, Report to CERN Conference (1962).

21. B. Lee and R. Sawyer, Preprint, Princeton Institute of Advanced Study (1962).

22. D. Amati *et al.*, *Phys. Letters* **1**: 29 (1962); and Report to CERN Conference (1962).

23. M. Gell-Mann and B.M. Udgoankar, *Phys. Rev. Letters* **8**: 346 (1962).

On Regge Poles in Perturbation Theory and Weak Interactions

K. RAMAN

MATSCIENCE
Madras, India

---

1. INTRODUCTION

In perturbation theory, a composite state must appear as a sum over Feynman diagrams and, presumably, so must a Regge pole. Recently, work has been done which shows how Regge poles might be obtained by summing certain classes of Feynman diagrams.[1-4] Blankenbecler, Cook, and Goldberger[5] have speculated that the photon may be a Regge particle, and have discussed the observable consequences of this. Also, Levy[6] has suggested that the radiative corrections in electron scattering sum to give a Regge behavior.

We shall here discuss the possible role of Regge poles in weak interactions.

That the concept of Regge poles may prove useful in weak interactions is suggested by the following:

1. Composite-state models have been useful in providing a theory of weak interactions that links these with the strongly interacting particles.[7,8] In particular, certain observed regularities of weak interactions, e.g., the conservation of the vector current and "partial conservation" of the axial-vector current seem to follow naturally from such composite-state models.

As a correct description of composite particles seems to be closely related to the idea of Regge poles, it may be of interest to examine the role of the latter in weak interactions.

2. Regge poles may help in solving the divergence problem of weak interactions. As is well known, neither the universal Fermi theory nor the charged vector boson theory gives a finite result when considered

with respect to all orders; also, the universal Fermi theory violated unitarity at high energies.[9] Thus, a theory that starts with the UFI or the vector boson interaction as the basic interaction requires a damping mechanism that gives a finite result.[10]

A somewhat similar situation also occurs in electrodynamics, where the lowest-order term in electron scattering seems to become unreasonably large at higher energies. (This is suggested by the observation by Blankenbecler, Cook, and Goldberger[5] that the one-photon exchange contribution to pp scattering exceeds the strong interaction contribution at a sufficiently high energy.) However, Levy[6] has pointed out that in electron scattering, the infrared radiative corrections provide a damping mechanism; further, the total effect (lowest order plus the infrared corrections) is equivalent to that of the positronium Regge pole in the crossed reaction. Thus, the positronium Regge pole provides a means of correctly summing the set of Feynman diagrams that dominates the high-energy behavior in electron scattering and gives a damped asymptotic behavior (as compared with one-photon exchange).

One may speculate that such a property is quite general, i.e., that although the weak interaction theories mentioned above are unrenormalizable, one may conjecture that a correct (field) theory of weak interactions would give a high-energy behavior that is equivalent to the contribution of a few Regge poles in the crossed reaction. Sections (1) and (2) suggest that we may attempt to use the idea of Regge poles as generalized composite states to incorporate the observed regularities of weak interactions in a finite theory. (It is known that the slope of the trajectories on the Chew–Frautschi plot may be related to the "size" or radius characterizing the scattering. In particular, a horizontal trajectory would correspond to point particles. There may be a similarity between an attempt to obtain a convergent theory using Regge poles and earlier attempts to obtain finite theories by introducing a "finite length" of Heisenberg.)

2. CURRENT–CURRENT INTERACTIONS AND REGGE POLES

The current–current picture of an interaction makes possible a factorization of the coupling strength in the lowest order in perturbation theory. When the interaction is iterated and summed to all orders, then the result (if formally obtainable) cannot be expected to be factorable into a similar form. However, if in some limit, e.g., that of high energy, this sum has a Regge behavior, then we can assume a factorization of the coupling $\beta(t)$ and think of the Regge trajectory as providing a generalization of the current–current interaction.

Concerning this Regge trajectory, we have different possibilities:

A. The trajectory may reduce to the original current–current interaction at the point $J = 1$ or $\alpha(y) = 1$. For weak interactions this may be looked upon as a "reggeization" of the weak current itself, analogous to the reggeization of the photon considered by Blankenbecler, Cook, and Goldberger.[5]

B. Alternatively, the point $J = 1$ on the trajectory may not correspond to the original interaction, as for the positronium trajectory summing the dominant infrared radiative corrections for high-energy electron scattering.[6] (The point $J = 1$ on the positronium trajectory does not correspond to the photon.) If a similar result obtains in high-energy "elastic" neutrino reactions, one may regard the particle corresponding to $J = 1$ on the Regge trajectory as a real or fictitious composite particle, e.g., as the vector boson appearing in certain theories.

C. In the weak interactions of baryons and mesons, the dominating Regge trajectories may be the same as those in strong interactions.

Consideration of (A) and (B) raises the question of whether in strong interactions also the Regge trajectories may be obtained by starting with a current–current interaction such as those in the gauge theories of strong interactions. That this is possible is suggested by Gell-Mann and Goldberger's[4] suggestion that the radiative corrections from the coupling with a neutral vector meson field reggeize the nucleon pole.

An immediate consequence of replacing the current–current interaction by a Regge trajectory is the violation of the "local action of lepton currents," which is consistent with the expectation that the Regge behavior is in some way obtained by summing higher-order terms. (The hypothesis of Regge behavior prescribes a particular mode of violation of the "local action" of lepton currents.) The high-energy differential cross section of a neutrino reaction $\nu + T \longrightarrow F + l$ is no longer a quadratic function of the neutrino momentum p_ν, the lepton energy ω_l, or $\cos \theta_{\nu F}$ (as it should be for a local lepton current[11]; it is asymptotically of the form $s^{2\alpha(t)-1}$.

Form factors become a function of the energy as well as the momentum transfer (as for a reggeized photon); their energy dependence is asymptotic mainly in a factor $s^{\alpha(t)-1}$.

Considering leptonic weak interactions, we expect that at low energies and for physical points on the Regge trajectories, the leptons must be coupled locally in pairs; we shall assume that leptons are always coupled in pairs $(l\,\nu)$ to a Regge trajectory. This would imply the absence of Regge poles in backward "elastic" scattering $\nu + T \rightarrow F + l$, etc; thus backward scattering may be expected to be negligible even at moderate energies.

Concerning the number of Regge trajectories required and their properties, the following hypothesis may be made:

We may conjecture that there exists a Regge trajectory $\alpha_W(t)$ (and the corresponding antiparticle trajectory) characteristic of the weak interactions (just as the photon trajectory is characteristic of the electromagnetic interactions). In order to explain the main features of leptonic weak interactions, that is, the possibility of $\Delta S = \Delta Q$ and $\Delta S = -\Delta Q$ currents, $\Delta S < 2$, etc., one may, for instance, require the trajectory to have the internal quantum numbers of the sextet of vector bosons recently postulated by Lee.[12] (On the other hand, these observed regularities could follow from the properties of the strong Regge trajectories that may dominate the weak interactions of baryons and mesons.) If we postulate that the trajectory has an even signature, then the vector boson (like the Pomeranchuk) is not a physical particle and will not be observed. The coupling $\beta(t)$ of the trajectory to all physical systems must presumably vanish at the point $t = t_0$ such that $\alpha(t_0) = 0$. Regarding the value $t = t_1$, where $\alpha(t_1) = 1$, the usual arguments for a heavy vector boson do not apply when the trajectory has an even signature. As the weak current seems to be a vector (and axial vector) current in low-momentum transfer reactions at low energy (such as neutron decay, etc.), we expect that t_1 is small. Noting that both the Pomeranchuk trajectory and the photon trajectory have $\alpha(0) = 1$, we may speculate that $\alpha_W(0) = 1$ also, and that strong, electromagnetic, and weak interactions each have a characteristic Regge trajectory with $\alpha(0) = 1$, and perhaps the same slope. The Pomeranchuk presumably has even signature and definite isospin I; the photon trajectory would have an odd signature and would be a schizon with $I = 0$ or 1, while the W trajectory would have an even signature and the schizoid behavior of Lee's vector bosons.

If the W trajectory corresponds to alternative (B) above, its slope may have a different order of magnitude from those of the Pomeranchuk and positronium trajectories (which themselves differ in order of magnitude).[5] Regarding the parity of the W trajectory, we must presumably require that the "coupling" of the trajectory to leptons does not conserve parity (all along the trajectory). In the weak inter-actions of baryons and mesons, assumption (C) would have definite consequences. (The occurrence of $\Delta S = 0$ and $\Delta S = 1$ interactions and the absence of $\Delta S \geqslant 2$ interactions may be traced to the fact that any $S \geq 2$ mesonic states that may exist must presumably have a large mass and hence a low trajectory.)

1. One expects that all the Regge trajectories that appear in strong interactions would appear in weak reactions also. (Note: This is already expected of the physical particles on the trajectories in the different "pole approximation" pictures.)

The Pomeranchuk trajectory would appear only in "inelastic" neutrino reactions of the type

$$\nu + T \longrightarrow (l^- + F^+) + T'$$

where T and T' are the initial and final target particles, and F^+ is a group of mesons emitted in the forward direction together with the final lepton. If the Pomeranchuk has both $\Delta S = 0$ and $\Delta S = 1$ couplings with leptons, this would result in the production of π's and K's respectively in the forward cone. Dominance of the Pomeranchuk trajectory would mean that the dominant high energy ν–reactions would have a forward cone with total charge zero.

2. Reactions like

$$\bar{\nu} + p \longrightarrow n + l^+$$
$$N^{*0} + l^+$$

would receive contributions from $I \geq 1$, $S = 0$ trajectories alone, reactions producing hyperons or hyperon isobars only from $S = \pm 1$ trajectories, and reactions of the form

$$\bar{\nu} + p \longrightarrow (l^+ \pi^-) + p$$
$$(l^+ k^-) + p$$

from the Pomeranchuk as well as $I \geq 1$, $S = 0$ trajectories. A factori-zation of the residue at the poles would lead to obvious relations of the form

$$\frac{\sigma(\bar{\nu}p \longrightarrow n + l^+)}{\sigma(\bar{\nu}p \longrightarrow N^{*0} + l^+)} = \frac{\sigma(np \longrightarrow pn)}{\sigma(np \longrightarrow pN^{*0})}$$

$$\frac{\sigma(\bar{\nu}p \longrightarrow Y^0 l^+)}{\sigma(\bar{\nu}p \longrightarrow Y^* l^+)} = \frac{\sigma(\pi p \longrightarrow Y^0 K^*)}{\sigma(\pi p \longrightarrow Y^* K^*)}$$

if we assume that the same Regge pole dominates in the weak and the corresponding strong interactions.

Variation of the Cross Section at High Energies

3. For the reactions $\bar{\nu} + p \rightarrow n + l^+$ and $\nu + n \rightarrow p + l^-$ the high-energy behavior of the cross section has been estimated by Cabibo and Gatto,[13] using the form factors obtained from the conserved current hypothesis. They find that the cross sections tend to constant limits at very high energies. However, if a Regge trajectory dominates the cross section at high energies, the cross sections must slowly decrease with increasing energy at high energies.

4. For inelastic processes like $\nu + N \rightarrow N + l + \pi$, a single pion exchange, with the final pion being produced at the weak vertex, gives a cross section that increases rapidly at high energies.[14]

It is not known how the corresponding Regge pole contribution can be calculated, but an extension of the conjecture for $2 \rightarrow 2$ reactions would lead us to expect an asymptotic decrease in the cross section.

Also, reactions like

$$\bar{\nu} + N \rightarrow (l^+ \pi^-) + N$$
$$\nu + p \rightarrow (l^- \pi^+) + p$$

will presumably be dominated by the Pomeranchuk pole, and the reactions

$$\bar{\nu} + p \rightarrow (l^+ \pi^0) + n$$
$$\nu + n \rightarrow (l^- \pi^0) + p$$

would be comparatively suppressed, since they will receive contributions only from lower Regge trajectories.

Similar results may be expected for strangeness-changing weak reactions.

With any hypothesis about the type of Regge trajectories occurring, we may compare the predictions of the UFI theory and the Regge pole hypothesis.

For the reactions

$$\bar{\nu}_e + e^- \rightarrow \bar{\nu}_\mu + \mu^- \tag{a}$$

and

$$\nu_\mu + e^- \rightarrow \nu_e + \mu^- \tag{b}$$

UFI would give differential cross sections that become equal in the forward directions.[10] The allowed Regge trajectories are the same for the two reactions, since they are determined by the quantum numbers of the t-reaction; thus we expect the forward differential cross sections for the two reactions to behave similarly here also. Concerning the backward scattering, we note that (a) and (b) are the crossed reactions of each other. If v_μ and v_e have the same leptonic number L, then each side of (a) has $L = 0$, while (b) has $L = 2$. Also, the muon number n_μ of (a) is 0 while that of (b) is 1, assuming additive muon number conservation. Thus, Regge trajectories contributing to (a) must have $L = 2$, $n_\mu = 1$, while those contributing to (b) must have $L = 0$, $n_\mu = 0$; hence trajectories similar to the π, K, etc., could contribute to (b) but not to (a). The backward scattering would be quite different for the two reactions.

3. REGGE POLES IN LOW-ENERGY WEAK INTERACTIONS

There is no reason to assume that the Regge pole contributions should dominate at low energies. However, as mentioned above, replacing the amplitude by a Regge pole may be regarded as an improved "pole approximation," in that the pole now includes the exchange of a set of particles with the allowed quantum numbers.

As any deviation from local action seems to be negligible at low energies, we assume an effective local vector or axial vector current coupled to the Regge poles in the "strong" form factors in meson or baryon decays. The contribution of a Regge trajectory to the weak form factor will be of the form

$$\frac{c(t)}{1 - \alpha(t)}$$

as for the electromagnetic form factors.[15] Approximating $\alpha(t)$ by a linear form gives back the "pole approximation" of dispersion theory; taking quadratic and higher terms gives deviations from the pole approximation. Application of this to meson and baryon decays is being studied. A difference between this and the electromagnetic case is that here there is an axial vector current also; thus trajectories corresponding to axial-vector mesons (like the one suggested in Dennery and Primakoff[16]) and possible 1^+ strange mesonic states will contribute. As in Freund's treatment of the two-photon exchange,[17] one may also take the coupling of the weak current (acting twice) to the Pomeranchuk trajectory.

4. CONSERVED CURRENTS AND REGGE TRAJECTORIES

If we assume the existence of the photon Regge trajectory and a weak vector trajectory, then the conservation of the corresponding currents implies that the coupling strength $\beta(t)$ is universal at the point t_1 such that $\alpha(t_1) = 1$. A possible generalization that immediately suggests itself is that such a universality may hold all along the "conserved" Regge trajectory, which would then be characterized not only by a definite shape $\alpha(t)$ but also by a definite residue $\beta(t)$.

If the photon trajectory has this property, it may be interpreted as a generalization of current conservation to an S-matrix theory of electrodynamics. (However, this is still a symmetry imposed *ad hoc* and not derived.) One result that such a conjecture would imply is that the electromagnetic form factors of all particles are the same. It would be of interest to test this experimentally. A similar assumption that the weak trajectory with negative parity is "conserved" would lead to the universality of all weak vector form factors. This would give a definite relation between K decay and Λ, Σ decay, π decay, and n decay, etc.

Partial conservation of the axial vector current, if assumed to be true, may find its analog in the smoothness properties of $\alpha(t)$ and $\beta(t)$.

5. COMPOSITE-PARTICLE MODELS AND REGGE TRAJECTORIES

The fact that summing over a set of Feynman diagrams can give a Regge behavior suggests an equivalence between conventional field theory and S-matrix theory with Regge poles. It may be useful to attempt to get this equivalence explicitly in a composite model theory.

We may start with the "minimal model" of Okun and take, for instance, the Λ, n, and p as the basic fields. One may attempt to obtain all the Regge trajectories appearing in strong interactions as composite states formed from Λ, n, and p. The fact that ideas such as conserved and partially conserved currents follow naturally from a Sakata model suggests, as pointed out above, that the same composite states may determine weak interactions also. Apart from the Λ, n, and p fields it is presumably necessary to introduce the μ, e, ν_μ, and ν_e fields also in order to include the weak and electromagnetic interactions as well.

6. REGGE POLES FROM PERTURBATION THEORY IN WEAK INTERACTIONS

Finally, there is the question of obtaining a Regge behavior explicitly by summing suitable sets of Feynman diagrams. Owing to the non-renormalizability of both the UFI and the vector boson interactions, it is not clear how this can be done explicitly. One may conjecture that the ladder diagrams may add here also to give a Regge behavior. If vector bosons exist, their emission may give an exponential factor at high energies (as with the infrared corrections in electron scattering). The emission of neutral vector mesons presumably gives an exponential factor; the question of proving this also for charged vector bosons is being studied.

One may also think of the leptons as being reggeized by the vector boson radiative corrections (as the nucleon would be); however, this would presumably be much smaller than the reggeizing effect of the electromagnetic radiative corrections (for the charged leptons), and the idea of regarding the neutrino as a Regge pole does not seem to be of direct utility.

NOTE

Feynman has observed that the principle of minimal electromagnetic interaction (which excludes terms like Pauli moments in the inter-action) restricts the high-energy behavior of cross sections; the presence of a Pauli moment, for example, would mean more subtractions. (See Feynman, Report to the Solvay Congress, 1961.) He has also suggested that one may try to replace the principle of minimal electro-magnetic interaction by a statement that high-energy electrodynamic cross sections are limited in some particular way.

We note that an appropriate replacement of the principle of minimal electromagnetic interaction may be the requirement that the high-energy cross sections have a Regge behavior. We may conjecture a similar principle for weak cross sections to replace the idea of a minimal weak interaction.

Restrictions on the high-energy behavior of cross sections may also be expressed by imposing γ_5-invariance; this may also be related to a Regge type of behavior, since it limits the number of subtractions (e.g., it excludes Pauli-moment terms).

REFERENCES

1. B. Arbusov *et al.*, *Phys. Rev. Letters* **4**: 272 (1962).
2. L. Bertocchi, S. Fubini, and M. Tonin, *Nuovo Cimento* **25**: 626 (1962).
3. J. C. Polkinghorne, *High-Energy Behaviour in Perturbation Theory* (preprint), University of Cambridge, 1962.
4. M. Gell-Mann and M. L. Goldberger, *Phys. Rev. Letters* **9**: 275 (1962).
5. R. Blankenbecler, L. F. Cook, and M. L. Goldberger, *Phys. Rev. Letters* **8**: 463 (1962). (See also preprint.)
6. M. Levy, *Phys. Rev. Letters* **9**: 235 (1962).
7. L. B. Okun, Proceedings of the International High Energy Physics Conference, CERN, 1958, p. 223.
8. S. Okubo and R. E. Marshak, *Phys. Rev.* **123**: 382 (1961).
9. T. D. Lee and C. N. Yang, Lectures at CERN, 1961.
10. A. Pais, IAEA Seminar, Trieste, 1962.
11. A. Pais, *Phys. Rev. Letters* **9**: 117 (1962).
12. T. D. Lee, *Phys. Rev. Letters* **9**: 319 (1962).
13. N. Cabibbo and R. Gatto, *Nuovo Cimento* **15**: 304 (1960).
14. N. Cabibbo and da Prato, *Nuovo Cimento* **25**: 611 (1962).
15. M. McMillan and E. Predazzi, *Nuovo Cimento* **25**: 838 (1962).
16. P. Dennery and H. Primakoff, *Phys. Rev. Letters* **8**: 350 (1962)
17. P. G. O. Freund, *Nuovo Cimento* **26**: 633 (1962).

Determination of Spin-Parity of Resonances

G. Ramachandran

MATSCIENCE
Madras, India

This discussion is not meant to be a comprehensive survey of the methods that have been used or suggested for the spin-parity determination; rather, its scope is restricted to a brief description of the two principal methods, which are (1) the Adair method and (2) the Dalitz method.

The Adair method[1] was first suggested in 1955 for the determination of the Λ^0-spin from

$$\pi^- + p \longrightarrow K^0 + \Lambda^0$$
$$\swarrow \qquad \downarrow$$
$$2\pi \quad N + \pi$$

Let us denote by A, B the spins of the particles K^0 and Λ^0, respectively, and let the spins add up to form a total spin S. Let L denote the relative orbital angular momentum of the two particles and j, m the total angular momentum and projection quantum numbers (Fig. 1).

If the Λ^0 decays into a nucleon and a pion with relative orbital angular momentum l (i.e., in the rest frame of the Λ^0), l and the nucleon spin $\frac{1}{2}$ combine to form B. Since the pions do not carry intrinsic spin, K^0 decays into two pions with relative orbital angular momentum A. We can therefore write the amplitude for the process as

$$f = \sum_{\lambda} T_{\lambda} L(m)(2j+1)^{-1/2} \sum_{m_L} c(Lsj; m_L m_S m) Y_{L, m_L}(\theta, \phi)$$
$$\sum_{m_A} c(ABS; m_A m_B m_S) Y_{A, m_A}(\alpha, \phi_{\alpha}) \qquad (1)$$
$$\sum_{m_l} c(l\tfrac{1}{2}B; m_l \mu m_B) Y_{l, m_l}(\beta, \phi_{\beta})^{1/2}\chi_{\mu}$$

where λ denotes the angular momentum and parity channel quantum numbers $\lambda = \pi, L, S, j, m$; the π is the parity, $\pi = (-1)^{L+l+A+1}$; and the

$L(m)$ are unit vectors such that $L(m) L(m') = \delta_{mm'}$, the initial system being unpolarized. If we choose the direction of the incident beam as the axis of quantization, the value of m must be equal to the spin orientation $(\pm\frac{1}{2})$ of the initial proton, the projection of the initial orbital angular momentum on the beam direction being zero; the various angles are shown in Fig. 1.

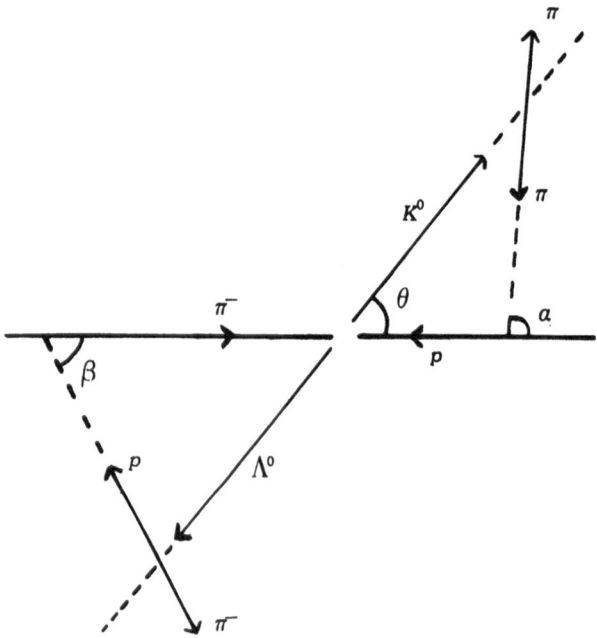

Fig. 1. The reaction $\pi^- + p \to \Lambda^0 + K^0$ in the center of momentum frame.

If we now select events in which the Λ^0 and K^0 are produced parallel to the beam, m_L has to be zero; and if we further restrict the study to the angular distribution of the Λ^0 decay products (in the Λ^0 rest frame) in conjunction with the K^0 decaying along or opposite the beam when $m_A = 0$ only contributes to (1) and $m_B = m$, the amplitude of interest reduces to the form

$$\sum_{m_l} c(l\,\tfrac{1}{2}\,B; m_l \mu m) Y_{l,\,m_l}(\beta, \phi_\beta)^{1/2} \chi_\mu \tag{2}$$

which results in angular distributions dependent on the Λ^0 spin B, as shown in Table I.

On the other hand, selecting again the events with production along the beam and studying the angular distribution of the K^0 decay products (in the K^0 rest frame with respect to the incident beam direction) in conjunction with the Λ^0 decaying along (or opposite) the beam, it is readily seen that we have now $m_A = m - \mu$ and consequently

$$Y_{A, \, m+1/2} \quad \text{and} \quad Y_{A, \, m-1/2} \tag{3}$$

contribute to the amplitude, resulting in an angular distribution of the form

$$a \cos^{2A} \alpha + a' \cos^{2(A-1)} \alpha \tag{4}$$

so that a study of the distribution determines A or if $a = 0$, at least $A - 1$, since a and a' cannot vanish simultaneously.

The application of the method to the determination of the K^* spin from

$$K^+ + p \rightarrow K^{*0} \; + \; N_{33}^{*++}$$
$$\downarrow \qquad\qquad \downarrow$$
$$K^+ + \pi^- \quad p + \pi^+$$

is theoretically straightforward, the K^{*0} with spin A decaying into two spinless particles and N_{33}^{*++} (with spin B) decaying into a nucleon and a pion. Recent experimental data[2] gave a good fit with a pure $\cos^2 \alpha$ distribution, thus indicating that the K^* spin A is >1, which combined with earlier evidence that A is < 2, determines the spin-parity of the K^* as 1^-. The experimental observations also favored the K^* spin to be strongly aligned in the $m = 0$ state with reference to the incident beam direction.

The second principal method, the Dalitz method,[3] is concerned with three-particle decays; for example, the ω decay into three pions

$$\omega \rightarrow \pi^+ + \pi^- + \pi^0 \qquad .$$

It is clear that in the rest frame of the ω^0, the three pions are coplanar and

$$m_\omega = T_0 + T_+ + T_- + 3m_\pi \tag{5}$$

where the T's denote the kinetic energies and m's the masses. If we now describe the decay events by plotting a point x for each event inside an equilateral triangle ABC such that the projections XP, XQ, XR are proportional to the kinetic energies T_0, T_+ and T_-, respectively, the requirement that these energies allow momentum conservation confines the points x to the inscribed circle within the triangle in the nonrelativistic case, the boundary distortion approaching a triangular shape in the extreme relativistic case (Fig. 2). The density distribution of points inside the boundary is characteristic of the spin-parity and other attributes of the decaying state. For example, since the isotopic spin of ω^0 is zero, we argue that the isotopic spins of any two pions must add up to one, and consequently the isotopic spin state of the system must be odd under exchange; since the pions are bosons the state must also be antisymmetric under space exchange. Therefore, for example, if the ω spin-parity is 0^- and if $\vec{q}$ denotes the relative momentum between any pair of pions and $\vec{p}$ the momentum of the third pion in the ω^0 rest frame, the matrix element for the decay is limited to scalars of the form

$$(\vec{p} \cdot \vec{q})F[(\vec{p} \cdot \vec{q})^2]$$

which vanishes when $\vec{p}$ is perpendicular to $\vec{q}$ or when the energies of any two of the pions are the same, that is, the density of points in the Dalitz plot must vanish along the lines AD, BE, and CF. In fact, since the space-wave function is antisymmetric, the density of points should be unchanged by a reflection across any of the three lines AD, BE, CF

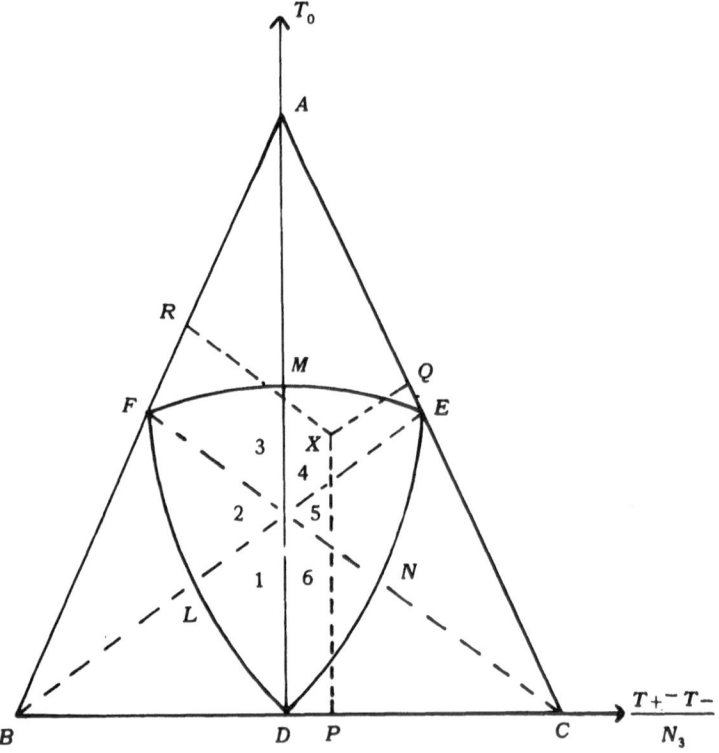

Fig. 2. The Dalitz plot.

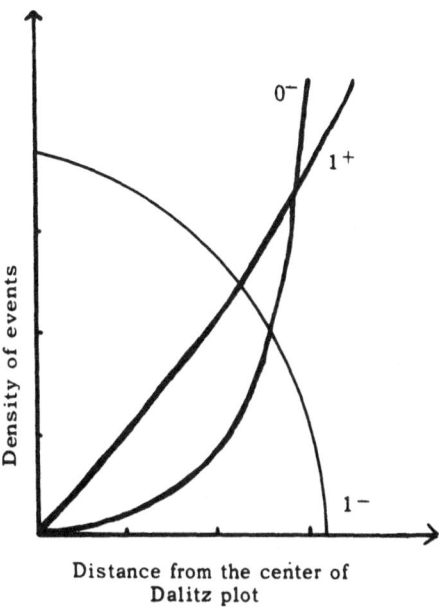

Fig. 3. Distribution of events in the Dalitz plot.

so that the distribution is essentially the same in each of the six sectors (irrespective of the spin-parity of the ω^0). The type of matrix elements M for various choices of spin-parity for the ω-meson and the corresponding features of the Dalitz plot are listed in Table II, and the dependence of the six folded $|M|^2$ on the distance from the center of the Dalitz plot is shown in Fig. 3. Experimental observations[4] favor the assignment 1^-, and since the simplest matrix elements for spin-parity assignments $2^\pm$ also vanish at the center of the plot, one concludes that the spin-parity of the ω-meson is 1^-. The assignment 0^+ is incompatible with 3π decay, since if we divide the 3π system into a system consisting of 2π's and a third pion and if l denotes the relative angular momentum between the two pions, the relative angular momentum between the two-pion and single-pion systems must also be l, so that the parity of the system must necessarily be odd, as are the pion pseudoscalar particles.

ADDENDUM

Recently, Charles Zemach[5] made a detailed study of the Dalitz plot for 3π decays, considering a large number of possible assignments of angular momentum, parity, and isotopic spin for the resonant states, speculating on the possible existence of such states from theoretical considerations on Regge recurrences and bootstrap mechanisms. To construct the appropriate decay matrix elements M, we proceed in a systematic way (which in summarized below) by writing the amplitude as a sum of products of the form

$$M = \sum M_I M_{JP} M_F \tag{1}$$

where M_I and M_{JP} incorporate, respectively, the dependence on isotopic spin, angular momentum, and parity (i.e., spherical tensors of rank I and J in the respective spaces) and the M_F are certain scalar functions (form factors) depending on the energies and momentum E_i and $\vec{p}_i$. Since the three pion momenta satisfy

$$\vec{p}_1 + \vec{p}_2 + \vec{p}_3 = 0 \tag{2}$$

in the rest frame of the parent particle, we have essentially two independent vectors and one independent axial vector

$$\vec{q} = \vec{p}_1 \times \vec{p}_2 = \vec{p}_2 \times \vec{p}_3 = \vec{p}_3 \times \vec{p}_1 \tag{3}$$

out of which M_{JP} has to be constructed; the pions being spinless particles, no spin operators are available. Further, the energies of the three particles satisfy

$$E_1 + E_2 + E_3 = m_x \tag{4}$$

m_x denoting the mass of the parent particle. The conservation laws (2) and (4) imply that any function of the energy and momentum variables could be expressed in terms of any two of them, and since we can express any scalar product $\vec{p}_i, \vec{p}_j$ in terms of the energies, for example

$$\vec{p}_1 \cdot \vec{p}_2 = \tfrac{1}{2}(E_3^2 - E_1^2 - E_2^2 - m^2) \tag{5}$$

the form factors M_F are essentially functions of energy only (in the barycentric frame).

Introducing the rectangular coordinates τ_x, τ_y, and τ_z in the isotopic spin space and observing that

$$\tau_\pm = \frac{\tau_x \pm i\tau_y}{\sqrt{2}} \tag{6}$$

$$\tau_0 = \tau_z$$

represent the three charge states of the pion, we can write scalar and vector products in isotpic spin space in the form

$$\vec{\tau}_1 \cdot \vec{\tau}_2 = \tau_0(1)\tau_0(2) + \tau_+(1)\tau_-(2) + \tau_-(1)\tau_+(2) \tag{7}$$

and

$$(\vec{\tau}_1 \times \vec{\tau}_2) \cdot \vec{\tau}_3 = i \begin{vmatrix} \tau_+(1) & \tau_+(2) & \tau_+(3) \\ \tau_-(1) & \tau_-(2) & \tau_-(3) \\ \tau_0(1) & \tau_0(2) & \tau_0(3) \end{vmatrix} \qquad (8)$$

Let E and O denote, generally, functions of energy momentum variables (i.e., M_{JP} and M_F put together) which are respectively completely symmetric and completely antisymmetric in the Particle Tables I, II, and III. Further, let A, B, C and $\tilde{A}$, $\tilde{B}$, $\tilde{C}$ denote general functions of the kinematic variables which have partial symmetry properties under exchange of the particle labels. If P_{ij} denotes the permutation operator, we define

$$\begin{aligned} P_{23} \quad A &= A & P_{23} \quad \tilde{A} &= -\tilde{A} \\ P_{13} \quad B &= B & P_{13} \quad \tilde{B} &= -\tilde{B} \\ P_{12} \quad C &= C & P_{12} \quad \tilde{C} &= -\tilde{C} \end{aligned} \qquad (9)$$

Since each of the pions carries isotopic spin 1, each $\tau(i)$ must occur to the first power in M. Further, since the pions are bosons, M must be completely symmetric under permutations $P_{ij} P_{ij}^{\tau}$ including isotopic

Table I

Spin value B	Angular distribution
$\frac{1}{2}$	Isotropic
$\frac{3}{2}$	$\frac{1}{2} + \frac{3}{2} \cos^2 \beta$
$\frac{5}{2}$	$\frac{3}{4} - \frac{3}{2} \cos^2 \beta + \frac{15}{4} \cos^4 \beta$

Table II

Spin-parity of ω°	Decay matrix element M	Density vanishes at
1⁻	$(\vec{p}_0 \times \vec{p}_+) + (\vec{p}_+ \times \vec{p}_-)$ $+ (\vec{p}_- \times \vec{p}_0)$	Whole boundary
0⁻	$(E_- - E_0)(E_0 - E_+)(E_+ - E_-)$	The straight lines AD, BE, CF of symmetry
1⁺	$E(\vec{p} - \vec{p}_+) + E_0(\vec{p}_+ - \vec{p}_-)$ $+ E_+(\vec{p}_- - \vec{p}_0)$	Center of the Dalitz plot

Table III. Structure of the Decay Matrix Elements M_{JP} of Types O, A, and E for Various Spin-Parity Assignments

Spin-parity	O	A	E
0^-	$s_{123} f_e$	f_1	f_e
1^+	$\overset{1}{\vec{p}_1}(f_2 - f_3) + \overset{1}{\vec{p}_2}(f_3 - f_1) + \overset{1}{\vec{p}_3}(f_1 - f_2)$	$\overset{1}{\vec{h}_{32}} p_2 + \overset{1}{\vec{h}_{23}} p_3$	$\overset{1}{\vec{f}_1} p_1 + \overset{1}{\vec{f}_2} p_2 + \overset{1}{\vec{f}_3} p_3$
2^-	$(s_2 - s_3) f_1 T(11) + (s_3 - s_1) f_2 T(22)$ $+ (s_1 - s_2) f_3 T(33)$	$f_1 T(11) + h_{32} T(22) + h_{23} T(33)$	$f_1 T(11) + f_2 T(22) + f_3 T(33)$
3^+	$(s_2 - s_3) f_1 T(111) + (s_3 - s_1) f_2 T(222)$ $\times (s_1 - s_2) f_3 T(333) + f_e\, O(3^+)$	$f_2 T(222) + f_3 T(333) + h_{23}$ $\times T(233) + h_{32} T(322)$	$f_1 T(111) + f_2 T(222) + f_3 T(333)$ $+ s_{123} f_e\, O(3^*)$
1^-	$f_e \overset{1}{\vec{q}}$	$(s_2 - s_3) f_1 \overset{1}{\vec{q}}$	$s_{123} f_e \overset{1}{\vec{q}}$
2^+	$f_1 T(1q) + f_2 T(2q) + f_3 T(3q)$	$h_{32} T(2q) - h_{23} T(3q)$	$(s_2 - s_3) f_1 T(1q) + (s_3 - s_1)$ $\times f_2 T(2q) + (s_1 - s_2) f_3 T(3q)$
3^-	$f_1 T(11q) + f_2 T(22q) + f_3 T(33q)$	$(s_2 - s_3) f_1 T(11q) + h_{32}$ $\times T(22q) - h_{23} T(33q)$	$(s_2 - s_3) f_1 T(11q) + (s_3 - s_2)$ $\times f_2 T(22q) + (s_1 - s_2) f_3 T(33q)$

Here $O(3^+) = T(112) - T(122) + T(223) - T(233) + T(331) - T(311)$.

spin space. Thus, for example, if the isotopic spin of the parent particle is zero, M can only have the form

$$M = \{\vec{\tau}(1) \times \vec{\tau}(2)\} \cdot \vec{\tau}(3) O \qquad (10)$$

This case has been considered at length in the present discussion. The above form (10) for M ensures immediately that the decay can proceed only through the charge mode, since otherwise the determinant (8) would vanish.

$I = 1$

To construct M in this case, we may combine the isotopic spins of a pair of pions to form a state with $I_{ij} = 0$, 1, or 2. The state with $I_{12} = 0$ has the form $(\vec{\tau}_1 \cdot \vec{\tau}_2)\vec{\tau}_3$, while the states with $I_{12} = 1$ or 2 are orthogonal mixtures of $(\vec{\tau}_2 \cdot \vec{\tau}_3)\vec{\tau}_1$ and $(\vec{\tau}_1 \cdot \vec{\tau}_3)\vec{\tau}_2$. The general amplitude M must therefore have the form

$$M = \vec{\tau}_1(\vec{\tau}_2 \cdot \vec{\tau}_3)A + \vec{\tau}_2(\vec{\tau}_3 \cdot \vec{\tau}_1)B + \vec{\tau}_3(\vec{\tau}_1 \cdot \vec{\tau}_2)C \qquad (11)$$

From the above expression we can immediately deduce, for example, the following branching ratios:

$$\frac{\Gamma(\pi^+ \pi^+ \pi^-)}{\Gamma(\pi^0 \pi^0 \pi^+)} = \frac{|A + B|^2}{|C|^2} \qquad (12)$$

$$\frac{\Gamma(\pi^+ \pi^- \pi^0)}{\Gamma(3\pi^0)} = \frac{|C|^2}{|A + B + C|^2} \qquad (13)$$

for the decay of a charged or neutral parent particle, neglecting, of course, any difference in phase space factors. Rewriting (12) in the form

$$\frac{\Gamma(\pi^+ \pi^+ \pi^-)}{\Gamma(\pi^0 \pi^0 \pi^+)} = \frac{4|A + B + C|^2 + |2C - A - B|^2}{|A + B + C|^2 + |2C - A - B|^2} \qquad (14)$$

we see that the branching ratio must lie between 4:1 and 1:1, even without knowing anything about A, B, C. In a similar way, it can be argued that (13) must either be equal to or greater than 2:3.

$I = 2$

Here, I_{12} can take the two values, 1 or 2, which correspond respectively to the antisymmetric and symmetric permutation properties under P_{12}^τ. The two isotopic spin states thus constructed may be denoted by $M_2^{(a)}$ and $M_2^{(s)}$. Under P_{13}^τ and P_{23}^τ, these states transform according to

$$
P_{13}^\tau \Longrightarrow
\begin{array}{cc}
(a) & (s) \\
\dfrac{1}{2} & \dfrac{\sqrt{3}}{2} \ (a) \\
\dfrac{\sqrt{3}}{2} & -\dfrac{1}{2} \ (s)
\end{array}
\qquad
P_{23}^\tau \Longrightarrow
\begin{array}{cc}
\dfrac{1}{2} & -\dfrac{\sqrt{3}}{2} \\
\dfrac{-\sqrt{3}}{2} & \dfrac{-1}{2}
\end{array}
\qquad (15)
$$

so that the general amplitude M may have either of the two forms

$$
M = M_2^{(a)}\sqrt{3}\,(A - B) + M_2^{(s)}(2C - A - B) \qquad (16a)
$$

$$
M = M_2^{(a)}(2\tilde{C} - \tilde{A} - \tilde{B}) + M_2^{(s)}\sqrt{3}\,(\tilde{A} - \tilde{B}) \qquad (16b)
$$

Considering, for instance, singly charged parent particles we can obtain

$$
\frac{\Gamma(\pi^+ \pi^+ \pi^-)}{\Gamma(\pi^0 \pi^0 \pi^+)} = \frac{|2C - A - B|^2}{|2C - A - B|^2} = 1 \qquad (17)
$$

while in the neutral case the $3\pi^0$ mode is forbidden.

To construct the factors M_{JP} with specified spin and parity of the parent state, we classify all the possible spin-parity assignments into two classes: (a) Normal parity states—states for which M_{JP} could be constructed out of $\vec{p}_1, \vec{p}_2, \vec{p}_3$ only; and (b) abnormal parity states—states for which M_{JP} involves $\vec{q}$ linearly—quadratic occurrences of $\vec{q}$ could be expressed again in terms of the $\vec{p}$'s. To construct spherical tensors of rank J, we choose J vectors $\vec{p}_a, \vec{p}_b \ldots \vec{p}_J$ or $\vec{p}_a, \vec{p}_b \ldots \vec{p}_{J-1}, \vec{q}$ (depending on whether the state is of normal or abnormal parity) and write the basic tensor

$$
T_{i_1, i_2 \ldots i_J} = (\vec{p}_a)_{i_1} (\vec{p}_b)_{i_2} \cdots \qquad (18)
$$

and subtract enough contracted terms to produce tracelessness in any pair of indices and then symmetrize it. For instance, some tensors for spin-2 are

$$
T_{ij}(11) = (\vec{p}_1)_i (\vec{p}_1)_j - \tfrac{1}{3}\delta_{ij} p_1^2
$$

$$
T_{ij}(1q) = \tfrac{1}{2}[(\vec{p}_1)_i (\vec{q})_j + (\vec{q})_i (\vec{p}_1)_j] \qquad (19)
$$

$$
T_{ij}(1,2) = \tfrac{1}{2}[(p_1)_i (p_2)_j + (p_2)_i (p_1)_j - \tfrac{1}{3}\delta(\vec{p}_1 \cdot \vec{p}_2)]
$$

We have already observed that the M_F are functions of the energies only. It is convenient to use instead of the E_i the variables s_i

$$
s_i = E_i - \tfrac{1}{3}(E_1 + E_2 + E_3) \qquad (20)
$$

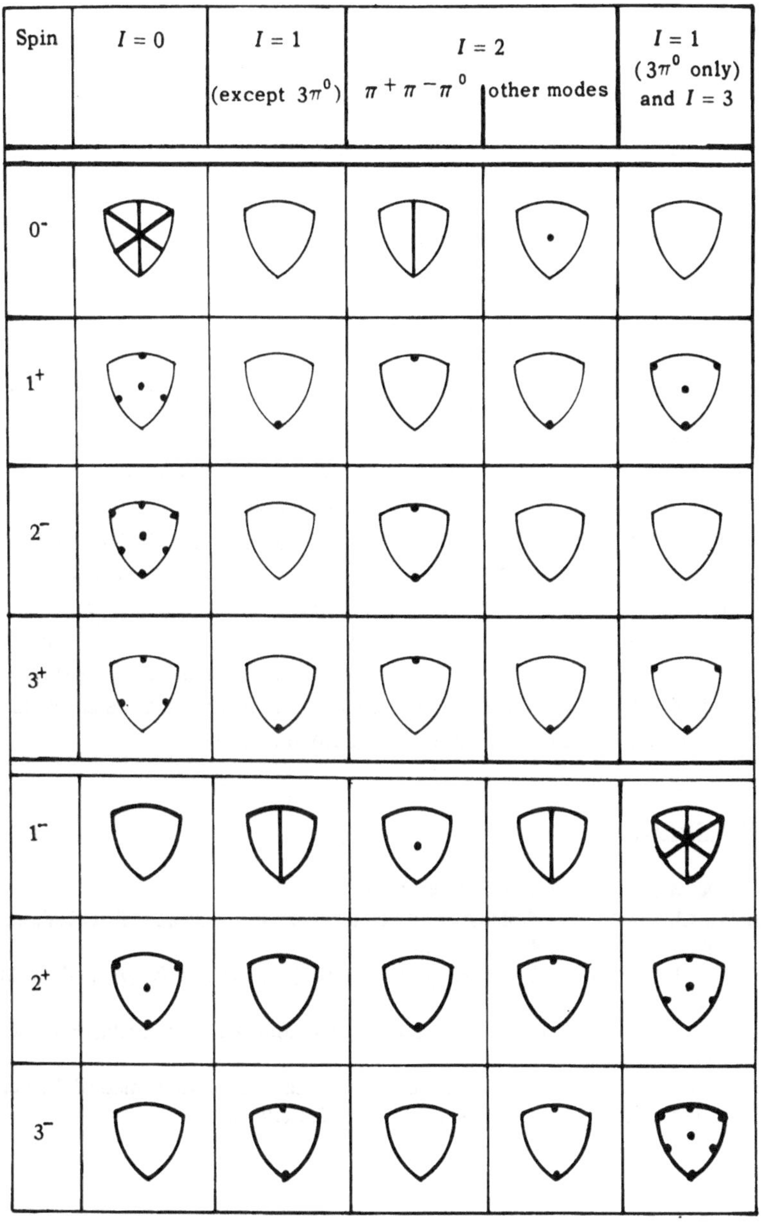

Fig. 4. Regions of the Dalitz plot for various spin-parity assignments where the density of events is expected to vanish as shown by thick lines or dots.

so that

$$\sum_i s_i = 0 \tag{21}$$

when $s_1 = s_2 = s_3$ marks the center of the Dalitz plot. The three meridians are described by $s_i = s_j$, and at the base of each meridian the third variable reaches the minimum value

$$s_{\min} = + \frac{Q}{3} \qquad Q = m_x - 3m_\pi \tag{22}$$

The maximum value is given by

$$s_{\max} = \frac{Q}{6_{m_x}} (m_x + 3m_\pi) \tag{23}$$

which lies between $Q/3$, the nonrelativistic limit, and $Q/6$, the relativistic limit.

Denoting by f_e a completely symmetric function of $s_1 s_2 s_3$, by f_i a function which is symmetric in the other two variables $j\,k$, by s_{123} a completely antisymmetric function, and by h_{ij} any arbitrary function $h(s_i, s_i)$, the general forms obtained for M_{JP} M_F for various spin-parity assignments belonging to the classes O, A, and E are given in Table III. Since the abnormal parity states are linear in $\vec{q}$, the plot density M^2 has a factor q^2, and consequently vanishes all along the periphery. The expected density distributions in the various cases have been worked out by Zemach and are shown in Fig. 4.

REFERENCES

1. R.K. Adair, *Phys. Rev.* **100**: 1540 (1955).
2. W. Chinowsky, G. Goldhaber, S. Goldhaber, W. Lee, and T. O'Halloran, *Phys. Rev. Letters* **9**: 330 (1962).
3. R.H. Dalitz, *Proc. Phys. Soc.* **A69**: 527 (1956). *Reports on Progr. in Phys.* **20**: 163 (1957).
4. M.L. Stevenson, L.W. Alvarez, B.C. Maglic, and A.H. Rosenfeld, *Phys. Rev.* **125**: 687 (1962).
5. C. Zemach, Lectures given at Matscience, The Institute of Mathematical Sciences, Madras, August 1963, *Phys. Rev.* **133B**: 1201 (1964).

Pion Resonances

T. S. SANTHANAM

MATSCIENCE
Madras, India

1. INTRODUCTION

As early as 1957, Nambu[1] remarked that it was rather difficult to understand such a large radius as that observed for the isoscalar part of the nucleon form factor and suggested the possibility that there might exist a meson coupled strongly with the nucleon which has the same quantum numbers as the photon. The simplest gauge-invariant coupling for this interaction is of the form

$$\lambda S_\mu \frac{\partial}{\partial X_\mu}(F_{\mu\nu}) = \lambda S_\mu \Box^2 A_\mu$$

using the Lorentz condition $\partial A_\nu/\partial X_\nu = 0$. As shown in Fig. 1, this

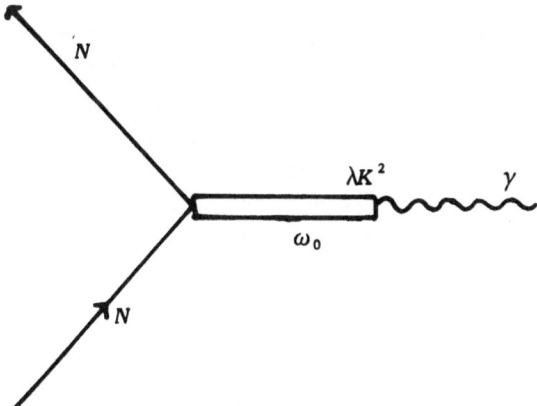

Fig. 1. Contribution to the form factors of the nucleon from vector meson exchanges.

interaction leads to the contribution $(-\lambda\,G\,K^2)/(K^2 - m_s^2)$ to the nucleon form factor; G is related to the strength of appropriate nucleon coupling and m_s is the mass of the meson. If the remaining contribution to this form factor is denoted by C (approximately independent of K^2, since they do not account for the large root mean square radius observed), the form factor is given by

$$F_s(K^2) = C - \frac{K^2}{K^2 - m_s^2} = F_s(0)\left[1 + \alpha - \frac{\alpha m_s^2}{m_s^2 - K^2}\right]$$

with $\alpha = \lambda G/C$. F is given by

$$F = F^{\mathrm{vec}} + F^{\mathrm{sc}}$$

The mean square radius is then

$$r^2 = (r^v)^2 + (r^s)^2 = 0.8\mathrm{f} \quad \text{for proton}$$
$$= 0.0\mathrm{f} \quad \text{for neutron}$$

Comparison of this with experimental data[2] for the charge and magnetic form factors requires the value

$$m_s \approx 4.8 m_\pi \approx 670\ \mathrm{MeV}$$

It was on this basis that Nambu suggested the existence of a massive $I = 0$ vector meson and discussed the decay process to be expected for this ω^0 state as

$$\omega^0 \longrightarrow \pi^+ + \pi^- + \pi^0$$
$$\longrightarrow \pi^0 + \gamma$$

through which it might be detected.

A similar possibility also existed for the interpretation of the iso-vector part of the nucleon form factor—that the observed isovector radius might be due to the existence of a massive $I = 1$ vector particle. The quantum number of such a particle allows it to decay rapidly into two pions, so that it would appear as a resonance in the p-wave pion–pion interaction. The existence of such a pion–pion resonance was suggested by number of workers; the details of its connection with the structure of the nuclear magnetic moment were first worked out by Frazer and Fulco.[3]

Also, from the Hofstader experiments, one concludes that

$$r^v \approx r^s \quad \text{and since} \quad r^2 = \frac{6}{m_{\mathrm{res}}^2}$$
$$m_{\mathrm{res}}^v \approx m_{\mathrm{res}}^s$$

Thus, the isoscalar particle of Nambu was expected to have nearly the same mass as the ρ predicted by Frazer and Fulco. Sudarshan predicted that

$$m_\rho < m_\omega < m_\rho + m_\pi$$

2. THE VARIOUS π-RESONANCES

The ρ-Meson

This particle was studied in the peripheral pion–nucleon collisions

$$\pi + N \longrightarrow 2\pi + N$$

and the photoproduction process

$$\gamma + N \longrightarrow 2\pi + N$$

(by the Pickup group,[4] the Anderson group,[5] and the Ervin group[6]). Its mass and width were found to be

$$m_\rho \approx 765 \text{ MeV} \qquad \Gamma/2 \approx 50 \text{ MeV}$$

It appears as a resonance in the pion–pion system in the $I = 1$, $J^{PG} = 1^{-+}$ state. The ρ-meson appears as a resonance in the pion–pion system. It is easy to see that the exchange of a ρ-meson between two pions yields an attractive force in the $J = 1, I = 1$ state. The dominant decay mode is

$$\rho^0 \longrightarrow \pi^+ + \pi^-$$
$$\rho^\pm \longrightarrow \pi^\pm + \pi^0$$

Some recent experiments seem to indicate a fine structure of the ρ_0-meson, and the original broad resonances seem to be ρ_1^0 and ρ_2^0, each of less than 10 MeV width.

The ω-Particle

The mass of the ω-particle is determined to be 787 MeV and $\Gamma/2 \leq 15$ MeV.

$$I = 0 \qquad J^{PG} = 1^{--}$$

Nambu and Sakurai[18] give the following decay rates with regard to the decay modes:

$$\Gamma_\omega \longrightarrow l^+ + l^- = 5 \times 10^{-4} \text{ MeV}$$
$$\Gamma_\omega \longrightarrow \pi^+ + \pi^- = 5 \times 10^{-3} \text{ MeV}$$

$$\Gamma_\omega \rightarrow \pi^+ + \pi^- + \pi^0 = 5 \times 10^{-2} \text{ MeV}$$

$$l = e \text{ or } \mu$$

The η^0-Meson

Pevsner *et al.*[7] observed two peaks in the histogram of the effective mass of the 3π system in the reactions

$$\pi^+ + d \rightarrow \underbrace{\pi^+ + \pi^- + \pi^0} + p + p$$

at 550 and 770 MeV. The larger peak near 770 MeV is clearly identifiable as ω_0. The other peak was identified as the η^0-meson. The mass and width were found to be $m_{\eta^0} = 546$ MeV, $\Gamma/2 \leqslant 25$ MeV, and $I = 0$. Bastien *et al.*[8] observed η^0 in the reactions

$$K^- + p \rightarrow \Lambda + \eta^0$$

$$\eta^0 \rightarrow \pi^+ + \pi^- + \pi^0$$

$$\rightarrow \text{neutrals}$$

They deduce the ratio

$$\frac{\eta^0 \text{charged}}{\eta^0 \text{neutral}} \text{ (at 760 MeV/}c) = 0.31 \pm 0.11$$

The assignment 0^{-+} is highly favored. The study of the decay models of η^0 probably takes up many pages of recent literature.

Pickup *et al.*[4] observe that

$$\sigma(\eta^0 \rightarrow \pi^+ + \pi^- + \pi^0) = 57 \pm 10 \ \mu\text{b}$$

It seems interesting and necessary to assume that the 3π decay mode (G forbidden for η^0) occurs through virtual electromagnetic transitions

$$\sigma(\eta^0 \rightarrow \text{neutrals}) = 140 \pm 55 \ \mu\text{b}$$

$$\frac{\Gamma_{\eta^0} \rightarrow \text{neutrals}}{\Gamma_{\eta^0} \rightarrow \pi^+ + \pi^- + \pi^0} = \frac{140 \pm 55}{57 \pm 10} \approx 2.5 \pm 1$$

Since the 3π decay mode could occur only through G-violating electromagnetic transitions,

$$\eta^0 \rightarrow \pi^+ + \pi^- + \gamma$$

should be relatively abundant.

But

$$\frac{\Gamma_{\eta^0} \rightarrow \pi^+ + \pi^- + \gamma}{\Gamma_{\eta^0} \rightarrow \pi^+ + \pi^- + \pi^0} < 9\% \left(\frac{5 \ \mu\text{b}}{55 \ \mu\text{b}}\right)$$

This result seems difficult to reconcile with $G = 1$ for η^0. Calculations give

$$\Gamma_{\eta^0} \rightarrow \pi^+ + \pi^- + \pi^0 = 1 \times 10^{-2} \text{ MeV}$$
$$\Gamma_{\eta^0} \rightarrow \pi^0 + \gamma = 3 \times 10^{-2} \text{ MeV}$$
$$\Gamma_{\eta^0} \rightarrow l^+ + l^- = 3 \times 10^{-4} \text{ MeV}$$
$$\Gamma_{\eta^0} \rightarrow \pi^+ + \pi^- = 0.5 \times 10^{-4} \text{ MeV}$$

Also

$$\eta^0 \rightarrow \pi^+ + \pi^- + \pi^0 \sim \alpha^2$$

(G-violating and hence through second-order electromagnetic interactions)

$$\eta^0 \rightarrow \pi^+ + \pi^- + \gamma \sim \alpha$$
$$\eta^0 \rightarrow \quad\quad 2\gamma \quad\quad \sim \alpha^2$$

A possible motivation on the basis of R-invariance was given by Marshak.

The ξ-Meson

The ξ-meson was obtained in the reaction

$$\pi^+ + p \rightarrow \pi^+ + \pi^0 + p$$

by Barlout et al.[9] at Saclay in France, and at Michigan, in the reaction

$$p + p \rightarrow d + \pi^+ + \pi^0$$

by Zorn[10] at Brookhaven, with $M_\xi = 575 \pm 15$ MeV, $\Gamma = 70$ MeV, and $I = 1$. Feld indicates that ξ could be a 0^{++} meson and that its dominant decay mode $\xi \rightarrow \pi^+ + \pi^0$ will also violate G-parity. But the existence of this particle has been questioned.

The *ABC* Particles

Abashion et al.[11] observed these in two events only; they are expected to be 0^+ with $I = 0$, if they exist.

The α-Meson

Pickup et al.[12] observed a peak in the reaction

$$p + p \rightarrow p + n + \pi^+ + \pi^+ + \pi^0$$

in addition to those in

$$p + p \rightarrow 2p + \pi^+ + \pi^- + \pi^0$$

But in the first reaction, the resonance at 625 MeV, $I = 1$, $\Gamma = 20$ MeV, is a 2π resonance. It is called the α–meson and is expected to be 1^+. The existence of this particle was predicted by Primakoff.

The χ-Meson

Lynch and Xuong[13] observed a peak in the 4π effective mass analysis in the reaction

$$p + \bar{p} \longrightarrow 3\pi^+ + 3\pi^-$$
$$\longrightarrow 3\pi^+ + 3\pi^- + 3\pi^0 \quad \text{at 1.05 BeV}$$

It is called the χ-meson.

The Chew-Theory Particle

Guiragossian et al.[14] observed a peak again at 1 BeV in the reaction

$$\pi^- + p \longrightarrow n + \pi^+ + \pi^-$$

This may be the spin-2 particle of Chew's theory.

The β-Meson

At $M = 420$ MeV, Schwartz et al.[15] observed a peak in the reaction

$$\pi^- + p \longrightarrow n + \pi^+ + \pi^-$$

called the β-meson.

Mechanism

Assuming the resonance to be dynamic, attempts were made to calculate the mass and coupling constant of the ρ-meson. But such attempts have not been successful because they involve the solution of a very complicated set of coupled integral equations on a computing machine. But the existence of the resonance seems approximately to follow from the operation of the "bootstrap mechanism," in which the strong force between two pions in a the p-state needed to produce the resonance, is provided by the exchange of a pair of resonating pions. Because of the inherent nature of this bootstrap mechanism, one would expect to be able to obtain the ρ-meson properties as the result of self-consistent calculation with no parameters.

The mechanism is applied on a simple scale with the approximation shown in Fig. 2. A ρ-meson of mass m_ρ coupled to the pion with a coupling constant $\gamma_{\rho\pi\pi}$ is exchanged between the two pions. The force thereby produced in the $J = 1$, $I = 1$ channel is attractive and depends on m_ρ and $\gamma_{\rho\pi\pi}$. In other words, the ρ-meson appears as a resonance in the $\pi-\pi$ system in the $J = 1$, $I = 1$ state. It is easy to see that the exchange of a ρ-meson between two pion yields an attractive force in the $J = 1$, $I = 1$ state. If the parameters of the ρ-meson are judiciously chosen, the attraction gives rise to a resonance whose mass and width are those assigned to the ρ-meson. The ρ-meson has therefore produced itself, so to speak.

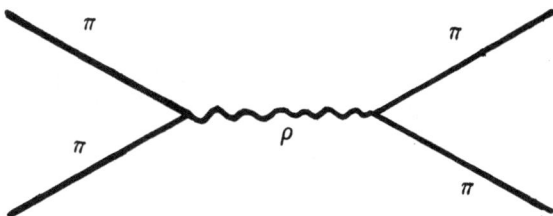

Fig. 2. Bootstrap of pions.

Thus we obtain two relations if m_ρ and $\gamma_{\rho\pi\pi}$ are adjusted suitably; this force may be made to produce a resonance at m_ρ with a coupling constant $\gamma_{\rho\pi\pi}$. From these two relations between these quantities we can determine both.

For a one-channel problem, the scattering amplitude $t(s)$ for a particular partial wave is a function of s, the total center of mass energy squared. It has a right-hand cut in the s-plane coming from its unitarity and from direct graphs (Fig. 3), and a left-hand cut due to the exchange graphs shown in Fig. 4.

If there are any stable one-particle states in this channel, $t(s)$ will have poles at the masses squared of such states. For s above threshold, the unitary condition on $t(s)$ states that

$$\frac{t(s) - t^*(s)}{2i} = t^*(s)t(s) \tag{1}$$

or

$$\operatorname{Im} t^{-1}(s) = -1$$

These analytic and unitary conditions are automatically satisfied

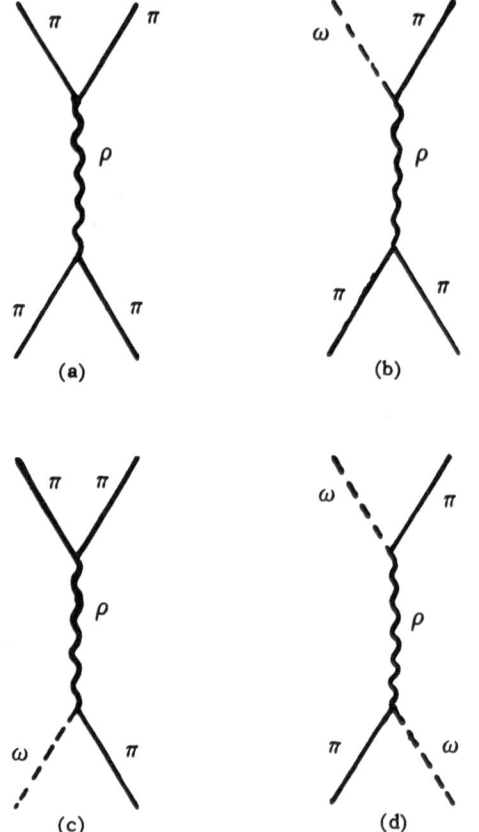

Fig. 3. Contribution to the right-hand cut from direct graphs.

if it is represented by

$$t(s) = -\frac{N(s)}{D(s)} \qquad (2)$$

where $D(s)$ has the same right-hand cut at t but no left-hand cut, while $N(s)$ has the same left-hand cut and with no right-hand cut and

$$D(s) = 1 + \frac{1}{\pi} \int\limits_{\text{threshold}}^{\infty} \frac{N(s')}{s' - s} \, ds' \qquad (3)$$

From the above three equations, N and D may be determined from their respective dispersion relations and the following expressions for absorptive parts:

$$\text{Im}\, D(s) = N(s)$$

$$\text{Im}\, N(s) = -D(s)\text{Im}\, t_L(s)$$

where $t_L(s)$ is the discontinuity of $t(s)$ across its left-hand cut. In the lowest order one would use $D = 1$, obtaining $N = -t_L$. Then N is obtained as an integral over the input t_L, so that finally we get

$$t(s) = t_L(s) \left[1 - \frac{1}{\pi} \int \frac{t_L(s')}{s' - s}\, ds' \right]^{-1}$$

For our case, the one-channel problem consists of forgetting the $\pi\omega$ channel and keeping only the $\pi\pi$ channel. The input is taken to be exchange of the ρ-meson between the two pions.

One may then compute the scattering amplitude $t(s)$ in the $J = 1$, $I = 1$ channel, in the simple approximation, and expect the meson itself to appear as a resonance with the same parameters as those used in the input diagram. It turns out that there is a self-consistent solution, and the resulting ρ-meson parameters are found to be $m_\rho = 350$ MeV, $\gamma^2_{\rho\pi\pi}/4\pi \approx 0.6$ Taking into account both the $\pi\pi$ and $\pi\omega$ channels, Zachariasen and Zemach[16] have worked out the mechanism. The most important deficiency of a calculation such as this lies in the violation of crossing symmetry.

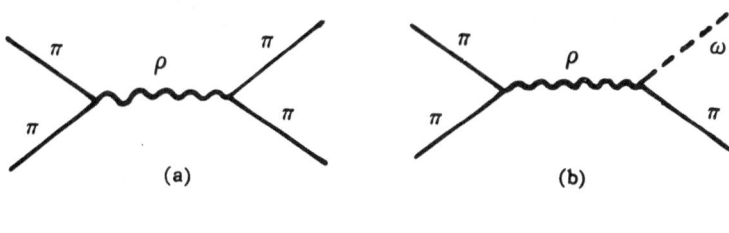

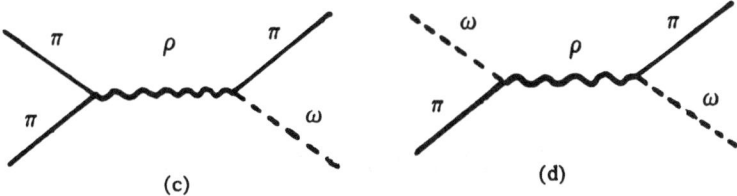

Fig. 4. Contribution to the left-hand cut from exchange graphs.

3. PARTICLE THEORY APPROACH TO 2π AND 3π SYSTEMS

This discussion is based on the work of Schiff.[17] We assume that 2π resonance arises from a potential and ask what effect this potential has on the 3π system. The potential method used here differs from field theory in three aspects: (1) The number of particles is fixed at 2π or 3π with no nucleon pairs. (2) A partially relativistic Schrödinger-type wave equation is used to describe the motion of these particles. (3) The interaction is represented by a static potential.

In the development of point (2), the wave function is assumed to depend on the average time of the several particles, not on the time differences. This means that the center of mass moves relativistically, but retardation effects are neglected in the internal motion. The justification is that retardation is less important in a strongly resonant or bound system in which the particles are close together most of the time.

A single free pion satisfies the Klein–Gordon equation*

$$(-\nabla^2 + m^2)\phi = -\frac{\partial^2 \phi}{\partial t^2}$$

where $\phi = \phi(\vec{r}, t)$ and m is the rest mass of the pion.

For two noninteracting pions, omitting isotopic spin dependence, an unsymmetrized wave function may be written as

$$\Psi(\vec{r}_1, t_1; \vec{r}_2, t_2) = \phi(\vec{r}_1, t_1)\, \phi(\vec{r}_2, t_2)$$

where each ϕ satisfies the Klein–Gordon equation, and Ψ satisfies the wave equation

$$(-\nabla_1^2 - \nabla_2^2 + 2m^2)\Psi = \left(-\frac{\partial^2}{\partial t_1^2} - \frac{\partial^2}{\partial t_2^2}\right)\Psi$$

Put

Average time	$t = \tfrac{1}{2}(t_1 + t_2)$
Relative time	$\tau = (t_1 - t_2)$
Center of mass coordinate	$\vec{R} = \tfrac{1}{2}(\vec{r}_1 + \vec{r}_2)$
Relative coordinate	$\vec{r} = \vec{r}_1 - \vec{r}_2$

Then we get

$$(-\tfrac{1}{2}\nabla_R^2 - 2\nabla_r^2 + 2m^2)\Psi = \left(-\tfrac{1}{2}\frac{\partial^2}{\partial t^2} - 2\frac{\partial^2}{\partial \tau^2}\right)\Psi$$

---

* Unit $\hbar = c$ is assumed.

We neglect the dependence of Ψ on τ and introduce a static potential V_2 through which two pions interact:

$$[-\tfrac{1}{2}\nabla_R^2 - 2\nabla_r^2 + 2m^2 + V_2(\vec{r})]\Psi(t, \vec{R}, \vec{r}) = -\tfrac{1}{2}\frac{\partial^2\Psi}{\partial t^2}$$

Let

$$\Psi = \chi(\vec{r}) \exp i(\vec{P}\cdot\vec{R} - Et)$$

where E = total energy and $\vec{P}$ = total momentum. Then

$$[-2\nabla_r^2 + 2m^2 + V_2(\vec{r})]\chi(\vec{r}) = \tfrac{1}{2}(E^2 - P^2)^{1/2}\chi(\vec{r})$$

This shows that the center of mass moves relativistically like a particle with rest mass $(E^2 - P^2)^{1/2} = M_2$. Thus, M_2 is the total internal energy of the system. The wave equation for the internal motion is written as

$$[-2\nabla_r^2 + V_2(\vec{r})]\chi(\vec{r}) = (\tfrac{1}{2}M_2^2 - 2m^2)\chi(\vec{r})$$

which is of the Schrödinger type.

The foregoing discussion can be extended to any number of pions.

REFERENCES

1. Y. Nambu, *Phys. Rev.* **106**: 1366 (1957).
2. Hofstadter *et al.*, *Phys. Rev.* **110**: 552 (1958); *Phys. Rev.* **111**: 934 (1958).
3. W.R. Frazer and J.R. Fulco, *Phys. Rev. Letters* **2**: 369 (1959).
4. Pickup *et al.*, *Bull. Am. Phys. Soc.* **6**: 301 (1961).
5. Anderson *et al.*, *Phys. Rev. Letters* **6**: 365 (1961).
6. Erwin *et al.*, *Phys. Rev. Letters* **6**: 624 (1961).
7. Pevsner *et al.*, *Phys. Rev. Letters* **7**: 421 (1961).
8. Bastien *et al.*, *Phys. Rev. Letters* **8**: 114 (1962).
9. R. Barlout *et al.*, *Phys. Rev. Letters* **8**: 326 (1962).
10. B.S. Zorn, *Phys. Rev. Letters* **8**: 282 (1962).
11. Abashian *et al.*, *Phys. Rev. Letters* **5**: 258 (1960).
12. Pickup *et al.*, *Phys. Rev. Letters* **8**: 329 (1962).
13. Lynch and Xuong, *Bull. Am. Phys. Soc.* **7**: 281 (1962); and N.Y. Xuong UCRL-10129 (Ph.D. Thesis).
14. Guiragossian, Powell, and White, *Bull. Am. Phys. Soc.* **1**: 281 (1962).
15. Schwartz, Kirt, and Tripp, *Bull. Am. Phys. Soc.* **7**: 282 (1962).
16. Zachariasen and Zemach, *Phys. Rev.* **128**: 849 (1962).
17. L.I. Schiff, *Phys. Rev.* **125**: 777 (1962).
18. J.J. Sakurai and Y. Nambu, *Phys. Rev. Letters* **8**: 79 (1962).

Pion-Nucleon Resonances

K. VENKATESAN

MATSCIENCE
Madras, India

The various pion–nucleon resonances experimentally known at present are given with relevant data in Table I.[1] In addition to these, a step in the $\pi^+ p$ cross section at approximately 800 MeV kinetic energy is known and a resonance in the $P_{1/2}$, $I = \frac{1}{2}$ state has been considered possible on the basis of the angular distribution at 900 MeV.

We shall study the mechanism of some of these resonances and the theories proposed in this connection. The earliest prediction of a series of isobars in the pion–nucleon system (before the pion was discovered) was the strong-coupling theory of Heisenberg, Wentzel, Pauli, and others.[2,3] One of the main features of this theory is that it takes into account the reaction of the meson field produced by the nucleon on the nucleon itself. This is a self-energy effect and, since the renormalization procedure was not then known, a finite size had to be assigned to the nucleon. Both the nucleon and the pion were treated classically, the latter because many virtual pions were expected to be closely associated with the nucleon. The Hamiltonian of interaction used here is the same as the one used later by Chew:

$$H_{\text{int}} = if_r \int U(\vec{r})\vec{\sigma} \cdot \vec{\nabla}\varphi \, d^3\vec{r} \tag{1}$$

where $U(\vec{r})$ is the source function of the nucleon. The most important result of the theory is that the nucleon will have excited or isobaric states with spin $J = \frac{3}{2}, \frac{5}{2}, \ldots$. The excitation energy E turns out to be proportional to $(J^2 - \frac{1}{4})$, but the proportionality constant which fits the N_{33}^* resonance does not lead to correct positions for the higehr resonances.

Chew's theory, which has in common with the above the idea of a

Table I

Resonances	Total pion energy (lab, MeV)	Mass (MeV)	Full width (MeV)	J	I	P
N_{33}^*	303	1237	94 ± 16	$\frac{3}{2}$	$\frac{3}{2}$	$+$
N_{13}^*	731	1515 ± 3	115 ± 25	$\frac{3}{2}$	$\frac{1}{2}$	
N_{15}^*	1020	1685 ± 5	100 ± 10	$\frac{5}{2}$	$\frac{1}{2}$	
N_{37}^*	1450	1920	200	$\frac{7}{2}$	$\frac{3}{2}$	
N_1^*	2080	2190 ± 20	200 ± 20		$\frac{1}{2}$	
N_3^*	2510	2360 ± 25	200 ± 25		$\frac{3}{2}$	

fixed extended source and hence a pseudovector interaction of the pion–nucleon system, also gives a good account of the $\frac{3}{2}, \frac{3}{2}$ resonance. Renormalization effects are incorporated. In the Chew–Low[4] version of the theory, the T-matrix element for pion–nucleon scattering which obeys the Low equation is T_1^2, given in terms of the phase shifts by the relation

$$t_{21}(z) = \frac{-4\pi v(q_1)v(q_2)}{(4\omega_1\omega_2)^{1/2}} \sum_{\alpha=1}^{4} P_\alpha(2,1) h_\alpha(z) \tag{2}$$

$$\lim_{\varepsilon \to 0} t_{21}(\omega_2 + i\epsilon) = T_1^2$$

Here (q_1, ω_1), (q_2, ω_2) are the momentum and energy of the initial and final pions, respectively; $v(q)$, the cut-off, is the Fourier transform of $U(r)$; and P_α are the projection operators for the four p-wave states corresponding to $J = \frac{1}{2}, \frac{3}{2}$, $I = \frac{1}{2}, \frac{3}{2}$. (But actually we need consider only three states, since $h_{31} = h_{13}$ in static theory. Thus $\alpha = 1, 2, 3$.)

$$\lim_{\varepsilon \to 0} h_\alpha(\omega_q + i\epsilon) = \frac{e^{i\delta_\alpha(\omega_q)} \sin \delta_\alpha(\omega_q)}{q^3 v^2(q)} \tag{3}$$

$h_\alpha(z)$ has the same properties as t_{21} itself, that is, it has a pole at the single-nucleon intermediate state and branch points at higher-particle intermediate states. It has the crossing symmetry

$$h_\alpha(z) = \sum_{\beta=1}^{3} A_{\alpha\beta} h_\beta(-z) \tag{4}$$

Defining a new function

$$\frac{1}{g_\alpha(z)} = \frac{z}{\lambda_\alpha} h_\alpha(z) \qquad \lambda_3 = \tfrac{4}{3} f^2 \tag{5}$$

(f being the pseudovector pion–nucleon coupling constant) and making the one-meson approximation (i.e., retaining up to only the one-meson plus one-nucleon intermediate state) in the unitarity condition, the analytic properties of $g_\alpha(z)$, the unitarity condition

$$\text{Im } g_\alpha(\omega) = -\frac{\lambda_\alpha}{\omega} q^3 v^2(q) \qquad 1 < \omega < 2 \qquad (6)$$

and the crossing relation

$$\frac{1}{g_\alpha(z)} = \sum_\beta B_{\alpha\beta} \frac{1}{g_\beta(-z)} \qquad (7)$$

enable us to write

$$g_\alpha(z) = 1 - \frac{z}{\pi} \int_1^\infty d\omega_q \frac{q^3 v^2(q)}{\omega_q^2} \left\{ \frac{\lambda_\alpha}{\omega_q - z} + \frac{H_\alpha(\omega_q)}{\omega_q + z} \right\} \qquad (8)$$

where H_α is connected to the discontinuity across the left-hand cut. Modifying the matrix $B_{\alpha\beta}$, Chew and Low got a solution $g'_\alpha(z)$, which at the origin can be expanded in a Taylor series:

$$g'_\alpha(z) = 1 - r'_\alpha(z) + P'_\alpha(z^2) + 0(z^3) \qquad (9)$$

where

$$r'_\alpha = \begin{pmatrix} -1 \\ 0 \\ 1 \end{pmatrix} r'_3$$

which shows that if the effective range for the (33) state is positive, the effective range for the (11) state is negative and zero for the (13) or (31) state. Assuming the integral in equation (9) to be a slowly varying function of the energy ω_q, the real part of $g_\alpha(z)$ for the 33 state can be written as

$$\text{Re } g_\alpha(\omega_q) = \frac{\lambda_\alpha}{\omega_q} q^3 v^2(q) \cot \delta_\alpha(\omega_q) = 1 - r_3 \omega_q \qquad (10)$$

which develops a resonance when $r_3 = (1/\omega_q)$. With the same parameters, that is, the coupling constant f ($f^2 \approx 0.8$) and a cut-off approximately equal to 6, Chew's theory also explained the resonance in photoproduction. Application of relativistic dispersion relations has also explained the low-energy features of pion–nucleon scattering.[5,6]

The explanation of the higher resonances is rendered difficult by the inelastic channels which open up and compete with elastic scattering. The work of Peierls,[7] Goebel and Schnitzer,[8] Carruthers,[10] Ball and Frazer,[15] and Cook and Lee[9] shows how the opening of a production channel excites the elastic channel by unitarity, thereby giving rise to a resonance in the scattering process. "Cusps" or "rounded steps" appear in the S-matrix element for a process at the threshold for a new reaction. According to Nauenberg and Pais,[12] there is the possibility of these cusps becoming "wooly." The effects they consider are associated with the opening of the production channel consisting of a nucleon and an unstable vector meson which subsequently decays strongly into two pions. Cook and Lee[9] have shown that the inelastic channels can cause a resonance in the elastic channel even below the threshold of the inelastic channels if there is a strong coupling between the elastic and inelastic channels.

Peierls[7] assumes formation of the N_{33}^* isobar as a possible explanation of the large rise in the production cross sections which can in turn influence the elastic scattering cross section. If one of the two pions in the final state of pion production in pion–nucleon collisions is in a p-state and the other in an s-state relative to the nucleon we may have a situation corresponding to N_{13}^* resonance. If, however, both mesons are in a p-state with respect to the nucleon (of course, this statement has meaning only if the nucleon is static), then they must necessarily be parallel to each other so that the total angular momentum of the state is $\frac{5}{2}$, corresponding to N_{15}^*. But the same argument would give an isotopic spin of $\frac{5}{2}$ for the state.

If we consider the pion–pion resonance in the $T = 1$, $J = 1$ state as being the cause of enhancement in the inelastic process of pion

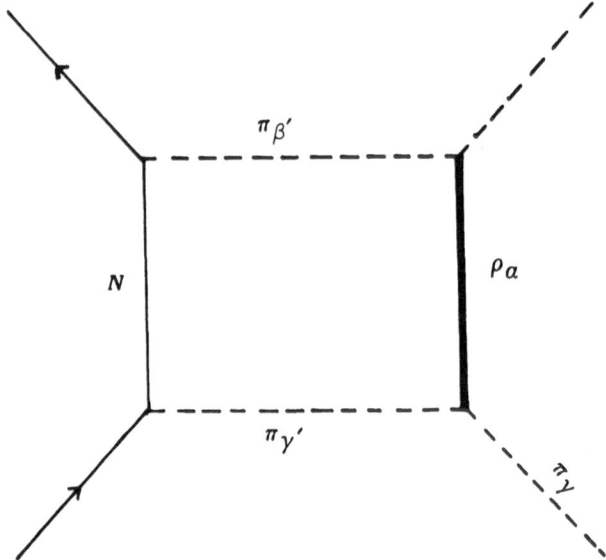

Fig. 1. Two-pion exchange graph.

production in pion–nucleon collisions ($\sigma = 43$ mb at 610 MeV) and hence indirectly the cause of the higher resonances in the elastic cross-sections, this difficulty is avoided. Itabashi et al.[13] assume a ρ–$\pi\pi$ interaction of the form

$$H_{\rho\pi\pi} = F \sum_{\alpha,\beta,\gamma} \chi_i^\gamma \partial_i \varphi^\alpha \varphi^\beta \epsilon_{\alpha\beta\gamma} \tag{11}$$

where α, β, γ refer to the isospin indices, and χ and φ are the ρ and π field operators, respectively. Similarly, the interaction among the ρ, π, and the electromagnetic fields is written

$$H_{\rho\pi\gamma} = iE \sum_{\substack{i,j,k,l \\ \alpha}} \epsilon_{ijkl}(\partial_i A_j - \partial_j A_i)\partial_k \varphi^\alpha \chi_l^\alpha \tag{12}$$

where A_i is the electromagnetic potential. The second and third resonances are fitted with a value of $F^2/4\pi \approx 5$. The observed second resonance in photoproduction is supposed to be due to the excitation of the resonant ρ–N state through a ρ-particle production by the incoming γ-ray in the pion field of the target nucleon. They show that the strong $\rho\pi\pi$ coupling and the interaction $H_{\rho\pi\pi}$ give a strong attractive potential of the π–N system in $I = \frac{1}{2}$ state for center of mass energy approximately equal to m_ρ. At these high energies, the large observed cross-sections can be expected only through a long-range pion–nucleon interaction. The longest range potential between a pion and nucleon is the two-pion exchange potential. The scattering amplitude corresponding to Fig. 1 becomes very large for $W \simeq m_\rho$, since energies of the intermediate ρN states can be very near to the initial energy W_i of the colliding system. Further, the pion–nucleon interaction becomes attractive, since the intermediate states have an energy larger than W_i.

The potential in the lowest-order perturbation theory is given by

$$V = \sum_n \frac{\langle f|H|n\rangle\langle n|H|i\rangle}{W_i - W_n}$$
$$= \sum_n \frac{|<n|H|i>|^2}{W_i - W_n} \tag{13}$$

if we take $i = f$. The sign of $W_i - W_n$ determines the sign of V. The initial energy W_i is less than the intermediate energy W_n for π^+–p scattering and $W_i > W_n$ for π^-–p scattering in the p-wave state. Therefore, the former is attractive and the latter repulsive. Similarly, the π–N force through the ρ–N intermediate states is attractive for $W_i > m_\rho$, since $W_n \geq m_\rho > W_i$ always. Even when W_i is greater than

m_ρ, as in the case of N_{15}^*, the potential is attractive, since most of the intermediate ρ–N states have $W_n > W_i$ when W_i is near to m_ρ.

The isotopic spin dependence of the π–N potential is given by

$$\langle \pi_\beta | V | \pi_\gamma \rangle \sim \sum_{\alpha',\beta',\gamma'} \epsilon_{\alpha\beta\beta'} T_{\beta'} \epsilon_{\alpha\gamma\gamma'} T_{\gamma'}$$

$$= 2\delta_{\beta\gamma} + \tfrac{1}{2}[\tau_\beta, \tau_\gamma]$$

$$= 4P_{1/2} + P_{3/2} \tag{14}$$

where P_I is the projection operator for isospin state I. The attractive π–N potential is four times as large in the $I = \tfrac{1}{2}$ state as in the $I = \tfrac{3}{2}$ state.

The angular momentum and parity of the resonances are fixed by considering the $\pi N \longrightarrow \rho N$ transition (Fig. 2). The matrix element in the lowest-order perturbation is

$$\langle \rho^\alpha, q | T | \gamma, k \rangle = \sum_{\gamma'} \epsilon_{\alpha\gamma\gamma'} \gamma_{\gamma'} Ff$$

$$\times \frac{(\vec{\sigma} \cdot \vec{k} - \vec{q})[(\vec{\rho} \cdot 2\vec{k} - \vec{q}) - \rho_0(2k_0 - q_0)]}{[(\vec{k} - \vec{q})^2 - (k_0 - q_0)^2 + 1]} \tag{15}$$

where $(\vec{\rho}, \rho_0)$ is the polarization vector of the ρ-particle, $(\vec{k}, k_0)$ and $(\vec{q}, q_0)$ the four energy–momentum vectors of the pion and ρ–meson, respectively. For $q = 0$ the matrix element is proportional to

$$2Ff\frac{(\vec{\sigma} \cdot \vec{k})(\vec{\rho} \cdot \vec{k})}{m_\rho^2} \approx \frac{-4\pi}{(3)^{1/2}} \times 2Ff$$

$$\times [P(s_{1/2} \longrightarrow s_{1/2}) + P(d_{3/2} \longrightarrow s_{3/2})] \tag{16}$$

Similarly, retaining the part of the amplitude linear in $\vec{q}$, the ratio $f_{5/2}:p_{3/2}:p_{1/2} = 1:\tfrac{3}{8}:3.75$, which shows that in addition to the known $f_{5/2}$ resonance at 900 MeV, there is the possibility of a strong resonance in the $P_{1/2}$ state.

Fig. 2. Lowest-order graph for $\pi N \to \rho N$ transition.

The ratio of the amplitudes in the $I = 0$ and $I = 2$ can be argued out even with an effective s-wave π-π interaction $\frac{1}{4}\lambda\,(\varphi_\alpha\,\varphi_\alpha)^2$. The ratio is 5:2, which leads to a ratio of the pion production cross-section in the $I = \frac{1}{2}$ and $I_{3/2}$ states of 5:2. If one of the final pions and the nucleon resonate in a relative $I = \frac{3}{2}$ state, the ratio becomes 10:1, thus favoring the $I = \frac{1}{2}$ state heavily.

Carruthers[10] and Goebel and Schnitzer[8] use the Chew–Low method to study the pion production process in pion–nucleon collision. They consider the π-π interaction to be the dominant one, the π-N p-wave interaction entering only as a final-state interaction. The chief graphs they consider are, therefore, Figs. 3 and 4. This leads to an integral equation the kernel of which is the π-N scattering matrix element, which is replaced by the resonant $(\frac{3}{2}, \frac{3}{2})$ interaction, the inhomogeneous term corresponding to Fig. 3. They find a strong enhancement in the partial transition $(D_{3/2} \longrightarrow P_{3/2}s)$ in the $I = \frac{1}{2}$ state, which is connected with the $I = \frac{1}{2}$, $D_{3/2}$ pion–nucleon resonance.

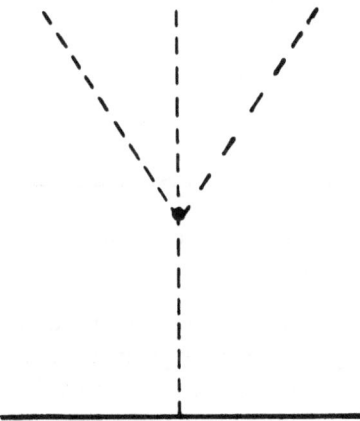

Fig. 3. Pion production in πN collision through π-π interaction.

Carruthers[14] suggests that the step observed at 850 MeV in the π^+-p system may possibly be a new resonance in the 850 to 950 MeV region, the parity and total angular momentum assignment being $D_{3/2}$. The isotopic spin is $\frac{3}{2}$. Since the direct and charge exchange cross-sections in the π^--p system are given in terms of the amplitudes for the $I = \frac{3}{2}$ and $I = \frac{1}{2}$ states by

$$\sigma_{\text{el}} \propto |f_{3/2}|^2 + 4|f_{1/2}|^2 + 4\,\mathrm{Re}(f_{1/2}^* f_{3/2})$$
$$\sigma_{\text{c.e.}} \propto 2|f_{3/2}|^2 + 2|f_{1/2}|^2 - 4\,\mathrm{Re}(f_{1/2}^* f_{3/2})$$

$$(17)$$

and since $f_{1/2}$ is known to vary in a resonant manner, the smallness and decrease of $\sigma_{c.e.}$ in the 500 to 800 MeV energy range requires that the interference term Re $(f_{1/2}^* f_{3/2})$ be large and positive. If the $I = \frac{3}{2}$ states were small and nonresonant, we should expect $\sigma_{c.e.}$ to rise abruptly on the high-energy side of the resonance.

Since $D_{1/2, \, 3/2}$ is the most important state in this energy interval, the simplest way to prevent both the occurrence of a peak in $\sigma_{c.e.}$ and an abrupt rise above the second resonance is to have the $D_{3/2, \, 3/2}$ phase shift positive and growing rapidly above 600 MeV.

As regards the photoproduction of pion pairs (enhancement in the cross section of which is likely to be significant for the higher resonances in single pion photoproduction), use of the Chew–Low method has

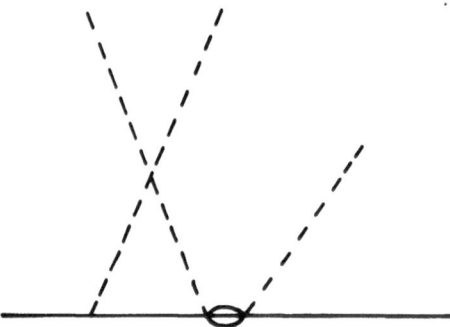

Fig. 4. Pion production in πN collision through π-π interaction and final state π-N p-wave interaction.

led us[11] to conclude that the increase in the total cross section in the 500 to 600 MeV region can be explained on the basis of the $(\frac{3}{2}, \frac{3}{2})$ isobar alone. The π–π interaction has a negligible role to play up to considerable values of the initial photon energy.

Finally, we shall review briefly the role resonances play in production amplitudes, using approximations to the Mandelstam representation for this purpose. Ball and Frazer[15] calculated the inelastic contribution to the higher partial waves by means of the strip approximation to the Mandelstam representation*[16] in which the principal

---

* The strip approximation maintains that the contributions to the physical amplitude from the double spectral functions, appearing in the Mandelstam representation, come mainly from the "elastic" strips of these functions adjoining the physical region. Thus, the contribution from Fig. 5 corresponds to that from the "elastic strip" of the t-channel in pion–nucleon scattering.

mechanism is the production of the $I = 1$, $J = 1$, π-π resonance. Figure 5 should give, according to the strip approximation, a correct estimate of the inelastic contribution for low-momentum transfer (high orbital angular momentum) scattering. Assigning to the π-π resonance a position and width in accordance with experiment, they find strong inelastic scattering in angular momentum and isospin states in which phenomenological analyses have suggested the presence of higher resonances in elastic scattering. The scattering amplitude in the physical region is $f(\nu) = [e^{i\delta(\nu)}\sin\delta(\nu)]/2i$ with $\delta(\nu)$ real; but above the inelastic threshold $\delta = \delta_R + i\delta_I$ where $\delta_I > 0$ according to unitarity. The inelastic cross section is calculated by assuming the "strip" approximation and projecting out partial waves from the matrix element. The inelastic cross section, which is given by

$$\sigma_I(\tfrac{3}{2}) = \frac{4\pi\,(J + \tfrac{1}{2})\,(1 - \eta^2)}{k^2}\frac{}{4} \tag{18}$$

rises rapidly in the region of the threshold for production of a π-π resonance, but rises to a height exceeding the unitarity limit by an order of magnitude. This is because the contribution of processes like $\pi + N \rightarrow \pi + \pi + N$ and $\pi + \pi + N \rightarrow \pi + \pi + N$ to the unitarity condition were not taken into account. Chew *et al.*[17] conjecture that the required unitarity damping in the inelastic part which comes about through the 4π, 6π, etc., contributions may be due to the fact

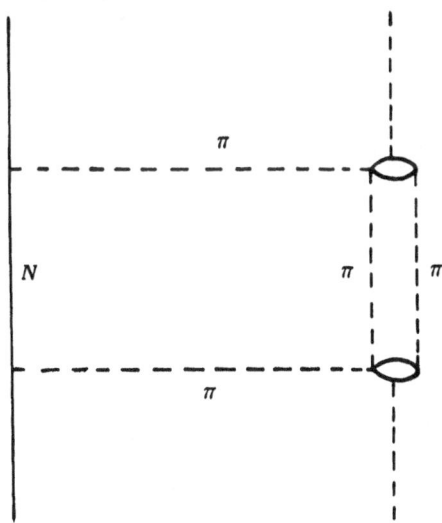

Fig. 5. Inelastic contribution to πN scattering in the strip approximation.

that such contributions appear as repulsive forces. In any case, the study of the multiparticle channels, which is a preliminary to their inclusion in the unitarity condition, can be done within the fold of the Mandelstam representation only if some pairs of resonating particles are considered as single particles. (In the π–N problem there is the π–π resonant state stimulated by the ρ-meson, the 3π state by the ω-meson and the π–N p-wave resonance by N_{33}^*.) Conversely, a study of interactions in which an unstable particle is present, such as $\pi + N \rightarrow \rho + N$, if ρ is considered an unstable particle, can be made only by considering processes such as $\pi + N \rightarrow \pi + \pi + N$, using the existence of the $\pi + \pi$ or $\pi + N$ resonance to reduce the complexity of the three-body state.

As an example of such a procedure let us consider the isobar approximation of Mandelstam *et al.*[18] p_1, p_2, p_3, p_4, and p_5 are the momenta of the particles in the process

$$1 + 2 \rightarrow 3 + 4 + 5$$

We can choose s_{12}, s_{13}, s_{15}, s_{34}, s_{23} as the independent invariants $[s_{ij} = (p_i + p_j)^2]$. s_{13} and s_{23} determine the direction of p_3 (or p_4). If particles 3 and 4 are resonating in an s-state, the corresponding amplitude is independent of s_{13} and s_{23}. Writing

$$\langle 4'3'|T|34\rangle = \frac{\beta}{M^2 - i\Delta - s_I} \tag{19}$$

$$s_I = s_{34}$$

where M is the mass of the isobar, Δ is connected with its width, and q is the momentum corresponding to the resonance, we can write the three-body amplitude as

$$\langle 543|T|12\rangle = F(s, t, s_I) \frac{\beta}{M^2 - i\Delta - s_I} + f_0(s_{12}, \ldots, s_{34}) \tag{20}$$

$$s = s_{12} \qquad t = s_{15} \qquad F(s, t, s_I) = F(s, t, M^2)$$

If we wish to include each pair of the three-particle final state as resonances we have to take a sum of three such F functions.

In the first approximation, we can neglect f_0 (In the model of Sternheimer and Lindenbaum, F is independent of t also.) Now $1 + 2 \rightarrow 1' + 2'$ will involve

$$\int <1'2'|T^*|345><345|T|12> d\tau \tag{21}$$

where $d\tau$ is over the phase space of 3, 4, 5 and this can be evaluated by substituting the above isobar approximation for the three-particle amplitude, and assuming that the region of phase space where two isobars can be formed simultaneously is small, i.e.,

$$\int F_k'^* F_l \, d\tau_{kl}$$

is neglected compared to

$$\int F_k'^* F_k \, d\tau$$

Mandelstam et al.[18] have used the above procedure in studying pion-nucleon scattering above the inelastic threshold.

REFERENCES

1. M. Roos, *Phys. Letters* **8**: 1 (1964).
2. A. Ramakrishnan, *Elementary Particles and Cosmic Rays*, Pergamon Press, Oxford (1962).
3. W. Pauli, *Meson Theory of Nuclear Forces*, Interscience Publishers, New York (1946).
4. G.F. Chew and F.E. Low, *Phys. Rev.* **101**: 1570, 1579 (1956).
5. G.F. Chew et al., *Phys. Rev.* **106**: 1337 (1957).
6. W.R. Frazer and J. Fulco, *Phys. Rev.* **117**: 1603 (1960).
7. R.F. Peierls, *Phys. Rev.* **118**: 325 (1960).
8. C.J. Goebel and H.J. Schnitzer, *Phys. Rev.* **123**: 1021 (1961).
9. L.F. Cook and B.W. Lee, *Phys. Rev.* **127**: 297 (1962).
10. P. Carruthers, *Ann. Phys.* **14**: 229 (1961).
11. S.K. Srinivasan and K. Venkatesan, *Nuclear Physics* **12**: 418 (1959); **29**: 335 (1962); **48**; 337 (1963).
12. M. Nauenberg and A. Pais, *Phys. Rev.* **126**: 360 (1962).
13. K. Itabashi et al., *Progr. Theoret. Phys.* **24**: 529 (1960).
14. P. Carruthers, *Phys. Rev. Letters* **4**: 303 (1960).
15. J.S. Ball and W.R. Frazer, *Phys. Rev. Letters* **7**: 204 (1961).
16. G.F. Chew and S.C. Frautschi, *Phys. Rev.* **123**: 1478 (1961).
17. G.F. Chew et al., *Phys. Rev.* **126**: 1202 (1962).
18. S. Mandelstam et al., *Ann. Phys.* **18**: 198 (1962).

The Influence of Pion-Nucleon Resonance on Elastic Scattering of Charged Pions by Deuterons

V. DEVANATHAN

UNIVERSITY OF MADRAS
Madras, India

Almost all experiments[1-3] reported in the literature are at low energy levels, for instance, 60, 85, and 140 MeV; the only experiment available at higher energy is that of Dul'kova et al.[4] Theoretical investigation[5-7] based on the impulse approximation accurately fits the experimental data at 60 and 85 MeV, but the agreement between theory and experiment seems to be poor at 140 MeV and the large theoretical cross sections at backward angles cannot be explained away by the on-the-energy-shell multiple scattering effects and the inclusion of s-wave phase shifts and D-state wave functions. As a result of this discrepancy at 140 MeV, Green[5] has made some skeptical remarks about the validity of the impulse approximation in the energy range of 140 MeV and above. Recently, Pendleton[8] has made an exhaustive study of the elastic scattering of charged pions at 142 MeV and the effect of the various corrections, using the form factor approximation. His final result does not seem to differ much from the numerical values we obtained earlier[7] by a simple approach using the Chew–Low amplitude for the scattering of pions by free nucleons.

We here present the cross section for the elastic scattering of charged pions at 300 MeV by deuterons. The details of the calculation have previously been outlined.[7] In Table I, the results of the present calculation along with two other sets of values obtained by others[1,9] are given for comparison with the experimental results. We obtained good agreement not only with regard to the angular distribution but also with the integrated cross section (Table II). In the present calculation, we do not distinguish between the scattering of π^+ and that of π^- by deuterons; the results will be the same for both.

V. Devanathan

Table I. Differential Cross Section $d\sigma/d\Omega$ in the Laboratory System for Elastic Scattering of Charged Pions from Deuterons (at 300 MeV in units of mb/sterad)

	Laboratory angle						
	0°	30°	60°	90°	120°	150°	180°
B.M. D.S.S. (Theoretical)	10.44	2.72	0.25	0.03	0.02	0.03	0.03
D.S.S. (Theoretical)	—	18.00	4.00	0	0	0	0
D.S.S. (Experimental)	—	7.5 ± 3	1 ± 0.5	1 ± 0.5	0.5 ± 0.25	0.25 ± 0.1	—
P.C.	22.79	7.954	0.8543	0.2007	0.2426	0.1785	—

B.M. D.S.S. (Theoretical) = The theoretical values of Bransden and Moorhouse[9] at energy 298 MeV.

D.S.S. (Theoretical) = The theoretical values at 300 MeV obtained on the impulse approximation.[1] The values are taken from the curve given for the purpose of comparison with the experimental results of Sachs and coworkers.

D.S.S. (Experimental) = The experimental values at 300 MeV taken from the experimental curve of Sachs and coworkers.[1]

P.C. = The present calculation at 300 MeV based on the impulse approximation using the Chew–Low amplitude or the scattering of pions from free nucleons.

It is to be emphasized that the experiments on elastic scattering are very important for testing the validity of the impulse approximation, since a reliable calculation can be made in this case because the final state of the system is well defined. Our investigation seems to indicate that the impulse approximation is a valid approximation in the low (85 MeV) as well as high energy state (300 MeV) but fails in the neighborhood of the pion–nucleon resonance. This conclusion[10] is based on the discordant results that we have obtained at 140 MeV. The continuance of experimental investigation at energies in the neighborhood of resonance (200 MeV) is strongly suggested, for it is hoped that these experiments will clearly decide the issue and, if our conjecture is confirmed, will stimulate the theoretical investigation of the influence of the pion–nucleon resonance on the impulse approximation. One plausible explanation is that in the resonance region, the pion–nucleon forms a quasi-bound state and consequently the interaction time is much longer, thus invalidating the impulse assumption that the other nucleon plays only the role of a spectator.

Table II. Integrated Cross Section in the Laboratory System for Elastic Scattering of Charged Pions by Deuterons (at 300 MeV in the angular range 15° to 170° in units of mb)

Experimental $\pi^+ + D \longrightarrow \pi^+ + D$	Experimental $\pi^- + D \longrightarrow \pi^- + D$	Theoretical
21 ± 6	14 ± 4	14

REFERENCES

1. A.M. Sachs, H. Wimick, and B.A. Wooten, *Phys. Rev.* **109**: 1733 (1958).
2. K.C. Rogers and L.M. Lederman, *Phys. Rev.* **105**: 247 (1957).
3. E. Arase, G. Goldhaber, and S. Goldhaber, *Phys. Rev.* **90**: 160 (1953). E.G. Pewitt *et al.*, Proceedings of the 1960 Annual International Conference on High Energy Physics, Rochester, New York, 1960, p. 196.
4. L.S. Dul'kova, I.B. Sokolova, and M.G. Shaframova, *Soviet Physics JETP* **8**: 217 (1959).
5. S. Fernbach, T.A. Green, and K.M. Watson, *Phys. Rev.* **90**: 161 (1953).
6. R.M. Rockmore, *Phys. Rev.* **105**: 256 (1957).
7. A. Ramakrishnan, V. Devanathan, and K. Venkatesan, *Nuclear Phys.* **29**: 680 (1962).
8. Pendleton, Private communication.
9. B.H. Bransden and R.G. Moorhouse, *Nuclear Phys.* **6**: 310 (1958).
10. V. Devanathan, *Nuclear Phys.* **43**: 684 (1963).

Pion-Hyperon Resonances

R. K. Umerjee

MATSCIENCE
Madras, India

1. INTRODUCTION

This is a study of the family of hyperon isobars characterized by $B = 1$, $S = -1$. The present situation is given in Table I.

Even from the table, it is clear that the intrinsic parameters of the resonant states such as spin, parity, and isotopic spin are not yet determined with any finality. These resonances appear most directly as resonances in the $\pi - Y$ system or in the $\bar{K}-N$ system, or in both.

The Y_1^* Resonance

The first such resonance to be established was the $\pi\Lambda$, or Y_1^*, resonance found by Alston et al.[1] in their study of the reaction

$$K^- + p \rightarrow \pi^+ + \pi^- + \Lambda \qquad (1)$$

for K^- laboratory momentum of 1150 MeV/c. The reaction is a two-step process, involving a $\pi\Lambda$ resonant state as an intermediate step, thus

$$K^- + p \rightarrow \begin{Bmatrix} Y_1^{*+} + \pi^- \\ Y_1^{*-} + \pi^+ \end{Bmatrix} \rightarrow \Lambda + \pi^+ + \pi^- \qquad (2)$$

The isotopic spin of this excited hyperon must be one, since it breaks up into a Λ and a π.

The Yale group[2] has observed Y_1^* in

$$K_2^0 + p \rightarrow \Lambda + \pi^+ + \pi^0$$

Dahl et al.[3] have presented evidence that Y_1^* plays a significant role in

$$K^- + d \rightarrow p + \pi^- + \Lambda$$

through the sequence

$$K^- + d \rightarrow p + Y_1^* \rightarrow p + \pi^- + \Lambda$$

Block et al.[4] have shown that Y_1^* production plays a similar role in the capture reaction $K^- + {}^4\text{He} \rightarrow \Lambda + \pi^- + {}^3\text{He}$.

The Y_0^* Resonance

The $I = 0$, Y_0^*, resonance at 1405 MeV was reported by Alston et al.[5] from $K^- \text{-} p$ reactions at $p_{k^-} = 1150$ MeV/c. Alston et al. studied the reactions

$$
\begin{aligned}
K^- + p &\rightarrow \Sigma^\pm + \pi^\mp + \pi^+ + \pi^- \\
&\rightarrow \Sigma^0 + \pi^0 + \pi^+ + \pi^- \\
&\rightarrow \Lambda + \pi^0 + \pi^+ + \pi^-
\end{aligned}
\tag{3}
$$

In May, 1962, the Alvarez group at Berkeley observed the

$$Y_0^* \quad \text{in} \quad K^- + p \rightarrow Y_0^* + \pi^0$$

for $p_{k^-} = 1.22$ BeV/c. The occurence of this resonance in reaction (3) at 1.22 and 1.53 BeV/c has also been confirmed by this group, and the observed width is 40 MeV.

The Y_0^{**} Resonance

The next $I = 0$ resonance, Y_0^{**}, at 1520 MeV has become established through $K^- \text{-} p$ collisions in the momentum range 300 to 500 MeV/c (as reported by Ferro-Luzzi et al.[6]).

The Y_0^{***} Resonance

A third $I = 0$, Y_0^{***}, resonance has been predicted[7] from a study of the total cross section for $K^- \text{-} p$ and $K^- \text{-} n$ collisions as functions of the $K \text{-} N$ total barycentric energy. The bump at 1810 MeV has been interpreted as an $I = 0$ resonance, Y_0^{***}.

The Y_2^* Resonance

Lundby et al.[8] have recently reported that a further Y^* resonance may exist in the mass region of Y_0^{***}. In the reaction

$$\pi^- + p \rightarrow K^+ + (X)^-$$

by plotting the K^+ production intensity as a function of the incident π^- momentum, they observed a third bump whose position corresponds to a mass value of 1550 MeV. (The first two peaks correspond to Σ^- production and Y_1^{*-} production.) If this does correspond to a Y^* resonance, its I-spin can only be 1 or 2. The absence of any evidence for an $I = 1$ resonance in K^-–p interaction for this mass value suggests that this bump might represent a Y_2^* resonant state. But in the CERN Conference (1962), Alston *et al.* observed that there is no significant peak in the 1550 MeV region for the interaction

$$K^- + p \rightarrow \Sigma^\pm + \pi^\pm + \pi^\mp + \pi^\mp$$

Hence, further evidence seems to be needed to establish this Y_2^* resonance.

Further Y^* Resonance

A further Y^* at 1685 MeV has been predicted by observation of the following:

1. A bump at 1685 MeV in the $\Lambda\pi$ effective mass distribution in $K^- + p \rightarrow Y + \pi$ at 1.5 GeV/c.
2. A bump at 1660 MeV in the $\Sigma\pi$ effective mass distribution in $\pi^- + p \rightarrow Y + K + \pi$ at 2.2 BeV/c.

Further evidence is needed to confirm this resonance.

2. SPIN-PARITY

We will first analyze the difficulties encountered in the determination of the intrinsic parameters of the resonant states. The most difficult situation is that where the resonance is relatively broad and experimentally accessible only as a final state interaction, as is the case with

Table I

State	Mass (MeV)	Width (MeV)	I-spin	Spin-parity
Y_1^*	1385	50	1	$(\frac{3}{2} +)$?
Y_0^*	1405	20	0	?
Y_0^{**}	1520	16	0	$(\frac{3}{2} +)$
Y_0^{***}	1815	>120	0	$> \frac{5}{2}$?
Y_2^*	1550	?	(2) ?	?

the Y^* resonances. The nonavailability of directly accessible entrance channels creates difficulties in the evaluation of the data about Y^* resonances, because the parameters can be studied only from their effects in the final-state interaction. In fact, even in the case of availability of an entrance channel, as in the case of the N^* resonance, there are many ambiguities in its interpretation until complete data become available.

There are certain interference effects which make it difficult to deduce the Y^* spin from the available data of the reaction

$$K^- + p \longrightarrow Y_1^* + \pi \longrightarrow (\Lambda + \pi) + \pi$$

Interference Effects†

Dynamic Interference.   The properties of the $\Lambda\pi$ resonance may be modified by interference with the primary pion in the above reaction. This would be particularly important close to threshold for Y_1^* production, where the primary pion moves slowly and is close to the Y_1^* system when it undergoes decay. At higher energies, the effect is less.

Interference Between Parallel Channels.   We have

$$K^- + p \longrightarrow \left\{ \begin{array}{c} Y_1^{*+} + \pi^- \\ Y_1^{*-} + \pi^+ \\ \Lambda + \rho^0 \end{array} \right\} \longrightarrow \Lambda + \pi^+ + \pi^- \qquad (4)$$

Even in the absence of dynamic interference, the amplitudes for the first two sequences must be taken coherently, as their interference gives rise to distortions of the angular distributions and resonance shape for the Y_1^* decay.

Interference with Background Production.   This refers to interference between the amplitudes describing the Y_1^* sequences (4) and nonresonant amplitudes which lead to the same final state $\Lambda\pi^+\pi^-$. It should be noted that the intensity of this nonresonant production need not be very large for interference to produce quite strong distortions from the distributions appropriate to the resonant production alone.

For the isolated model [which neglects the interference (a) between Y_1^* of opposite charge, (b) between Y_1^* and ($\Lambda\pi^+\pi^-$) produced in

---

† This section follows closely the lectures by Dalitz at Brookhaven National Laboratory: BNL 735 (T–264).

nonresonant background states, and (c) final-state interaction between π^+ and π^- and the influence of Bose statistics on this system], the Adair distributions gives 5-to-1 odds against $J = \frac{3}{2}$. But this is irrelevant because of the oversimplified nature of the model.

Y* Decay Angular Distribution with Respect to the Production Normal. If its spin were $\frac{1}{2}$, then in the absence of interfering background, the Y^* must necessarily decay isotropically with respect to production normal

$$\vec{n} = \frac{(\hat{K} \times \hat{Y}^*)}{|\hat{K} \times \hat{Y}^*|}$$

Ely *et al.*[9] have shown that the Y^* decay distribution is not isotropic, but fits the form $1 + a \cos^2 \theta$ with $a = 1.5 \pm 0.4$. This shows evidence for $J \geqslant \frac{3}{2}$. Further, even from the point of view of isotropic distributions, it is difficult to distinguish between $J = \frac{1}{2}$ and $J = \frac{3}{2}$, since Dalitz and Miller have shown that the requirements of Bose statistics on the final pions can cause a Y^* with $J = \frac{3}{2}$ to have a relatively isotropic angular distribution in the Adair analysis.

The most probable interpretations of the observed angular correlations in the Y_1^* decay require that the spin of Y_1^* be greater than $\frac{1}{2}$. If the spin of Y_1^* is then assumed to be $\frac{3}{2}$, the relative parity of the Y_1^* and Λ^0 must be even to account for the Λ^0 polarization. However, if the Y_1^* has spin $\frac{1}{2}$, the parity must be odd. From helium bubble chamber experiments, Block *et al.*[4] have strong evidence for the assignment $J = \frac{3}{2}$ for Y_1^*. If $J = \frac{3}{2}$, it is surmised that Y_1^* should be a $P_{3/2}$ state, from the copious yield of Y_1^*, as the angular momentum barrier should be lower for $L = 1$. But, still, the spark chamber data seem to favor spin $\frac{1}{2}$ for this resonance.

Beall *et al.* have obtained preliminary results on K^-–p elastic scattering cross section for $p_{k^-} = 700$ to 1400 MeV/c from spark chamber analysis. For Y_0^{***}, $\sigma_{k^--p}(\theta)$ elastic in the neighborhood of this resonance requires terms of $\cos \theta$ to the fifth power. This is consistent with the assignment $F_{5/2}$ to this resonant state. But data and analysis are too preliminary.

3. INTERPRETATION

Attempts have been made to inquire into the predictions of global symmetry regarding hyperon resonances. This affords a direct interpretation of the Y^* resonances as the analog of N^*. Following the prediction of Gell-Mann[10] that there are two resonances in the $P_{3/2}$ state with I-values 1 and 2 corresponding to N_3^* resonance, Amati et al.[11] have estimated the location of these resonances taking into account the $\Sigma\Lambda$ mass difference.

Kerth and Pais[12] have discussed a phenomenological extension of this analogy to the higher N^* resonances, on the basis of the global symmetry scheme. It is expected that an $I = 0$ and $I = 1$ Y^* resonance corresponds to each $I = \frac{1}{2}$ πN resonance and an $I = 1$, $I = 2$ Y^* resonance corresponds to each $I = \frac{3}{2}$ πN resonance.

Another familiar interpretation of Y_1^* is that it represents a virtual $\bar{K}N$ bound-state. But the recent analysis of Ross[13] and Humphrey,[14] as detailed by Hwa and Feldman, gives a modification of this interpretation.

The analyses by Ross and Humphrey of the low-energy K^-–p data render impossible the virtual $\bar{K}N$ bound-state interpretation of Y_1^*, although it is not ruled out for Y_0^*. By studying the resonances as manifestations of the dynamic structure of the closed pion–hyperon system, it is shown that a consistent interpretation of the observed resonances can be given.

Taking Y_0^*, since its mass is below the $\bar{K}N$ threshold, we need deal only with the $I = 0$ $\pi\Sigma$ channel. Upon examining the Born terms of the scattering amplitudes, it is readily seen that a resonance is possible in the $P_{3/2}$ state. For this state, the calculated half-width is 8.2 MeV, which can be compared with the experimental value of about 10 MeV. Evidently, it is possible to accommodate Y_0^* as a dynamic $\pi\Sigma$ resonance. With this interpretation for Y_0^* it is unlikely that Y_0^{**} is another resonance in the $P_{3/2}$ state having the same origin. By analogy with the π–N case we may expect it to be a higher resonance in the $D_{3/2}$ state of the $\pi\Sigma$ system.

A possible alternative to the above interpretation of the two $I = 0$ resonances is that we consider Y_0^* as a $\bar{K}N$ state (permitted according to solution II of Ross and Humphrey) and Y_0^{**} as a $\pi\Sigma$ dynamic resonance. In this scheme, Y_0^* is in the $S_{1/2}$ state of the $\pi\Sigma$ system and is unrelated to $P_{3/2}$ resonance, which can now accommodate Y_0^{**}.

The proper choice of these interpretations can be made only after more definite experimental information is available. Considering Y_2^*, we have a simple one-channel problem in the $I = 2$ state of the $\pi\Sigma$ system (ignoring three-particle channels). Further calculation then indicates that a resonance can exist in the $P_{3/2}$ state with a theoretically predicted half-width of 70 MeV. Verification of this resonance width for Y_2^* should be of great interest.

REFERENCES

1. M. Alston *et al.*, *Phys. Rev. Letters* **5**: 520 (1960).
2. H. Martin *et al.*, *Phys. Rev. Letters* **6**: 283 (1961).
3. O. Dahl *et al.*, *Phys. Rev. Letters* **6**: 142 (1961).
4. M. Block *et al.*, *Nuovo Cimento* **20**: 724 (1961).
5. M. Alston *et al.*, *Phys. Rev. Letters* **6**: 698 (1961).
6. M. Ferro-Luzzi *et al.*, *Phys. Rev. Letters* **8**: 28 (1962).
7. V. Cook *et al.* (unpublished), in: L. Kerth, *Rev. Mod. Phys.* **33**: 389 (1961).
8. J.D. Dowell *et al.*, Proceedings of the Aix-en-Provence International Conference on Elementary Particles, Vol. 1, CERN, Saclay, France, 1961, p. 385.
9. R. Ely *et al.*, *Phys. Rev. Letters* **7**: 461 (1961).
10. M. Gell-Mann, *Phys. Rev.* **106**: 1296 (1957).
11. D. Amati *et al.*, *Phys. Rev. Letters* **5**: 524 (1960).
12. L. Kerth and A. Pais, UCRL-9706.
13. R. Ross, UCRL-9749.
14. W. Humphrey, UCRL-9752.

Some Remarks on Recent Experimental Data and Techniques[†]

E. SEGRE

UNIVERSITY OF CALIFORNIA
Berkeley, California

1. INTRODUCTION

This paper is a brief account of some recent experimental results in elementary-particle physics, beginning with a discussion of resonances, mainly polypion, with mention of an experiment assigning the spin of the K^*.

The possible sources of information on the polypion resonances are the following:

1. Electron–positron colliding beam experiments.
2. Photoproduction of these resonances.
3. Nucleon form factors.
4. Peripheral pion–nucleon collisions, especially the reaction

$$\pi + N \longrightarrow 2\pi + N$$

5. $N\bar{N}$ annihilation experiments.

Experiment (1) is conceivable, but has not yet been carried out, and (2) also has not yet been extensively done.[‡] With regard to (3), attempts have been made to fit nucleon form factors with the resonances. The ρ-meson fits very well, but apparently not the ω-meson.

Most of the information, then, has come from (4) and (5). Method (4), that of peripheral pion–nucleon collisions is probably the best so

---

† This is based on the notes taken by T.K. Radha and K. Raman of an informal talk delivered by Prof. E. Segre at MATSCIENCE.

‡ However, there are double-photoproduction experiments which have shown the ρ-resonance; also, evidence for the *ABC* anomaly has recently been found in a photoproduction experiment.

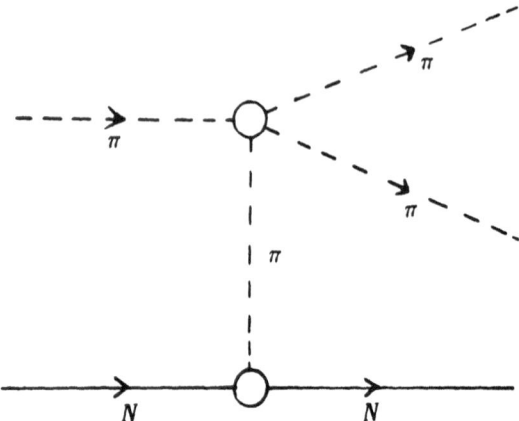

Fig. 1. One-pion exchange graph for double-pion production in pion-nucleon collisions.

far, because it is easy to get pion beams of different energies. The Chew–Low extrapolation method has been used in the study of peripheral pion–nucleon collisions leading to double pion production. The trouble with this method is that it gives an exact equality only in a limit (i.e., that the exchanged pion is on the mass shell); this limit cannot be attained, as it lies in the unphysical region (Fig. 1). (Measurement of effective-mass distributions and missing-mass distributions gives information on the mass and lifetime of the resonances; the extrapolation method gives the $\pi\pi$ cross section, according to the formula

$$\frac{\partial^2 \sigma}{\partial p^2 \partial \omega^2} = \frac{f^2}{2\pi} \frac{(p/2m)^2}{p^2 + m^2} \frac{\omega(\omega^2/4 - m^2)^{1/2}}{q^2} \sigma_{\pi\pi}$$

where the limit $p^2 \rightarrow m^2$ is to be taken.)

2. QUANTUM NUMBERS OF THE RESONANCES

The quantum numbers to be found are
(1) Mass (m)
(2) Lifetime (τ)
(3) Isospin (I)
(4) Spin (J)
(5) Parity (Π)
(6) G-parity (G), where relevant.

(1) The mass is perhaps the most easily fixed quantum number; it is given by the position of a peak.

(2) As the lifetime (τ) is very short, it cannot be measured directly. What is measured is the half–width (Γ), which is related to the lifetime by the uncertainty principle $\Gamma \approx \hbar/\tau$. The accuracy to which a width can be measured at present is of the order of 10 MeV. Thus the observed widths of all the resonances except the ρ-resonance are instrumental, as they are of the order of or less than 10 MeV.

(3) The isospin (I) can be determined fairly easily and unambiguously; this is done by observing in what charge-states the resonance occurs. For instance, if a resonance with zero strangeness occurs only in a state with total charge $Q = 0$ and not in $Q = 1, 2, \ldots$ states, then it may be assigned an isospin $I = 0$.

3. THE SPIN, PARITY, AND G-PARITY

The spin and parity of a resonance may be determined by two methods:

The Dalitz method

This may be used when the resonance is observed to decay into three particles. The observed decays are plotted on the triangular Dalitz plot; the distribution of points observed is compared with that expected for different spin assignments.

A requisite for the applicability of this method is that the resonance must decay purely through strong interactions. When electromagnetic interactions are also involved in the decay process, as in the decay of the η-meson, normal methods of analyzing the decay do not apply.

The G-parity of a resonance may also be obtained from the Dalitz plot if purely strong interactions are involved in the decay.

The Adair Method

Here the spin of a resonance is deduced from its two-body decay (into spinless particles) by observing the distribution of the decay products with respect to the momentum of the incident particle in the reaction in which the resonance is produced. Very recently, the spin of the K^* has been established to be 1^- by this method.

Table I gives some of the pseudoscalar mesons and resonances; the first row indicates that a PS-meson (0^-) has the same spin-parity as

Table I

	J^{PG}	J^{PG}
I	PS $\longleftrightarrow N\bar{N}$ in $^1S_0$ η 0^{-+}	V $\longleftrightarrow N\bar{N}$ in $^3S_1$ ω 1^{--}
0	$\longrightarrow 3\pi$ $\longrightarrow 2\gamma$	$\Rightarrow 3\pi$
$\frac{1}{2}$	K 0^- $\cdot\longrightarrow 3\pi$, etc.	K^* $1^{-\cdot}$ $\Rightarrow K + \pi$
1	π 0^{--} $\longrightarrow 2\pi$ $\cdot\longrightarrow \mu + \nu$	ρ 1^{-+} $\Rightarrow 2\pi$

Notation: $\Rightarrow$ denotes a strong decay.
            $\longrightarrow$ denotes a decay involving electromagnetic interactions.
            $\cdot\longrightarrow$ denotes a decay by weak interactions.

an $(N\bar{N})$ pair in a $^1S_0$ state, while a vector meson (1^-) has the same spin-parity as a $(N\bar{N})$ pair in a $^3S_1$ state. (One may think of the Fermi–Yang model where the pion is a bound $^3S_0$ state of a nucleon and antinucleon.) The K and K^* have, of course, a strangeness of ± 1; for them G-parity is not a useful concept. The η, K, and π cannot decay by purely strong interactions.

Three recent experiments relating to polypion resonances are:

1. The decay $\eta \longrightarrow 2\gamma$ has been observed recently at Frascati.[1]

2. A resonance at about 1250 MeV with $I = 0$ and decaying into 2π seems to have been observed. If confirmed, this could be the $J = 2$ resonance predicted by the Regge pole hypothesis (by a straight–line extrapolation of the Pomeranchuk trajectory).

3. The ABC anomaly has again been observed in a double–photo-production experiment at Frascati.[2] An up-to-date list of the various resonances has been given by A.H. Rosenfeld.[3] It should be noted that the existence of the ζ-meson cannot be considered as established.

Other data are: (a) The β-decay of the π-meson, i.e., the decay

$$\pi^+ \longrightarrow \pi^0 + e^+ + \nu$$

predicted by the Gell-Mann–Feynman theory, has been observed. It has a branching ratio of about 10^{-8}. (b) The two-neutrino hypothesis has received good support from a Brookhaven experiment, where it was observed that the neutrinos obtained from the decay mode $\pi \longrightarrow \mu + \nu$ produced muons but no electrons.[4] Further experiments on this are planned at CERN.

Finally, an important advance in experimental technique, the production of a target containing polarized hydrogen, has been made. This work was done by Chamberlain, Jeffries, Schultz, and Shapiro. The hydrogen nucleus in a sample of $La_2 Mg_3 (NO_3)_2 \cdot 24 H_2O$ with 1% of Mg replaced by a paramagnetic impurity (Nd^{142}) was polarized by a dynamic method. The sample was cooled to a temperature of 1.5°K in a magnetic field of 9169 G, and a microwave voltage at 34.3 kMc/sec was applied. The total mass of the sample was 20 g, containing 0.6 g of hydrogen. The electron–proton system has four levels, corresponding to the different spin orientations; the applied microwave voltage saturates one of the transitions. The polarization obtained was 20%. Only one proton in thirty was polarized; however, with coincidence methods (e.g., for *pp*-scattering), even this would be useful. (It is a free proton that is polarized. As scattering of an incident proton by a free target proton is characterized by a definite angle [$\theta = 90°$] between the final protons, it can be distinguished from the scattering on the unpolarized [bound] protons.)

This is only a beginning; a much larger degree of polarization is expected to be achieved with improvements. The availability of a polarized target would open up a world of new possibilities. It would be possible to do experiments in which a polarized proton beam is scattered off a polarized target of protons, which would directly give the spin dependence of phase shifts. Also, by observing the polarization of the Σ produced by π-mesons incident on polarized nucleons, the ($K\Sigma$) relative parity could be established with definiteness. The cross-section for this is given by

$$\sigma = \sigma_0(1 + \vec{P} \cdot \vec{P}_0)$$

where P_0 is the polarization of the target proton and P is the polarization of the Σ produced on an unpolarized target (at the same energy and angle). The two signs are for different relative parities. Various other experiments can also be carried out with polarized targets.

DISCUSSION

Question: Could you please explain the details of the experiment assigning the spin of the K^*?

Answer: The experiment is by W. Chinowsky, G. Goldhaber, S. Goldhaber, W. Lee, and T. O'Halloran. The reaction studied is

$$K^+ + p \rightarrow K^{*0} + N_{33}^{*++}$$
$$\downarrow \qquad\qquad \downarrow$$
$$(K^+ + \pi^-)\ (p + \pi^+)$$

All the particles produced are observed (in a bubble chamber). From the momenta of the final nucleon and pion, the momentum of the N_{33}^* in the center-of-mass system (and hence that of the K^*) is inferred. By selecting events in which the K^* is emitted close to the forward direction, the distribution in the angle between the final K^+ in the CMS of the K^* and the incident K^+ direction is observed (Fig. 2). The observed $\cos^2\alpha$ distribution is shown in Fig. 3; it indicates a spin $\geqslant 1$. Alston's earlier experiments showed $J < 2$; hence we conclude that the K^* spin is 1.

Question: What about the $(K\pi)$ resonance at 730 MeV?

Answer: According to Rosenfeld,[3] all the assignments for this resonance have question marks. Thus, the existence of this is doubtful.

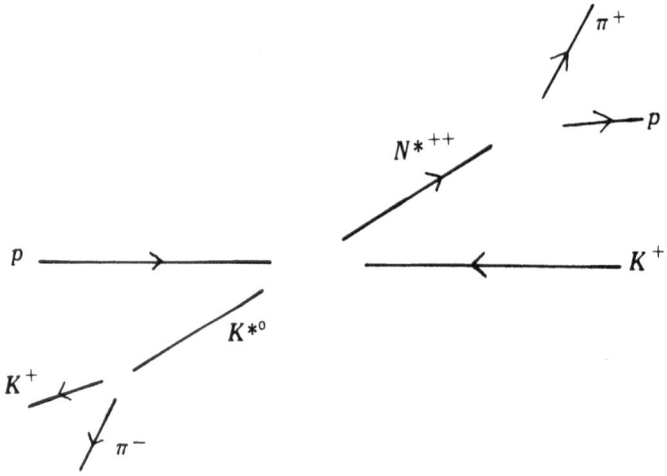

Fig. 2a. K^* production in K^+p collisions.

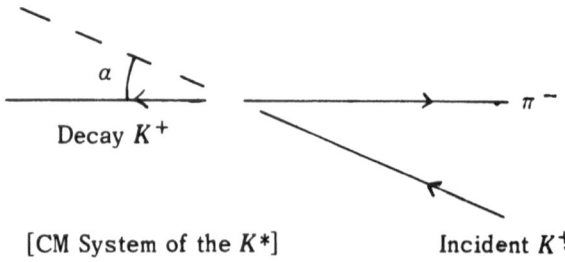

Fig. 2b. Orientations of the incident and decay K^+ in the center of mass of K^*.

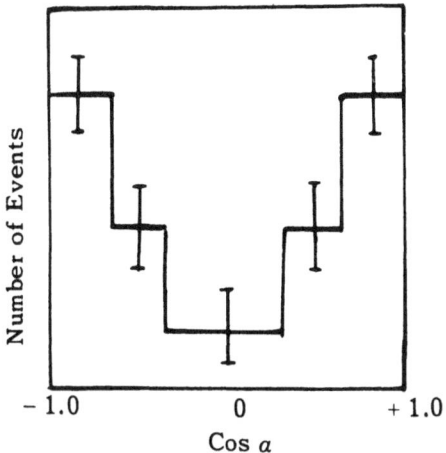

Fig. 3a. Cos α distribution in the production of K^*.

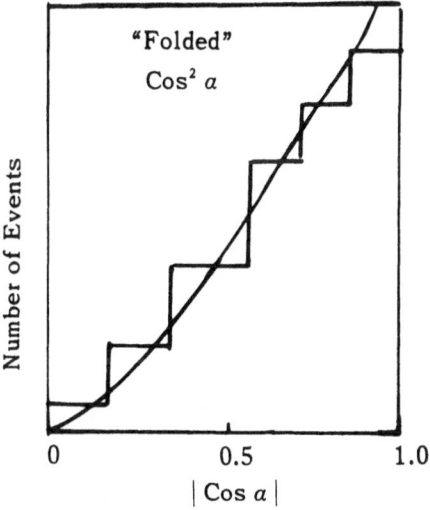

Fig. 3b. "Folded" $\cos^2 \alpha$ distribution.

REFERENCES

1. Mencuccini *et al.* (preprint).
2. B. Richter, *Phys. Rev. Letters* **9**: 217 (1962).
3. A.H. Rosenfeld, "Tentative Data on Strongly Interacting Particles," UCRL (September 1962).
4. G. Danby, *Phys. Rev. Letters* **9**: 36 (1962).

On New Resonances

BOGDAN MAGLIĆ

CERN
Geneva, Switzerland

---

1. INTRODUCTION

Generally, we think of the states of matter characterized by I-spin, spin J, parity, and a well-defined energy ($=$ mass) without making any distinction between resonances and particles. In particular, we shall deal here with bosons; such "states of matter" are called the "heavy mesons," or resonances.

One of the first indications that the resonances exist came from the nucleon–antinucleon annihilation experiments

$$p + \bar{p} \rightarrow N\pi$$

The statistical model predicts an average particle multiplicity ($\langle N \rangle$) of order 3. However, it was observed that

$$\langle N \rangle = 5.1 \pm 0.3$$

To reconcile this with the statistical model, it was suggested that the average $\langle N \rangle$ may be due to some three bodies in the final system rather than to three π-mesons, so that

$$p + \bar{p} \rightarrow \pi + \pi + (\pi\pi\pi)$$

or

$$p + \bar{p} \rightarrow \pi + (2\pi) + (2\pi)$$

which are in agreement with the experimental $\langle N\pi \rangle \approx 5$, and thus the existence of short-lived 3π or 2π particle states would reconcile the statistical model with experiment.

The ρ-meson was the first "resonance" observed. It was studied in the reaction[1-3]

129

$$\pi + N \longrightarrow \pi + \pi + N$$

and its mass and width were found to be

$$M_\rho \approx 765 \text{ MeV}$$
$$\Gamma = 100 \text{ MeV}$$

and it was assigned the quantum numbers

$$I = 1 \qquad J = 1^-$$

It was first predicted by Frazer and Fulco[4] to explain the isovector form factor of the nucleon. The dominant decay mode is

$$\rho^0 \longrightarrow \pi^+ + \pi^-$$
$$\rho^\pm \longrightarrow \pi^\pm + \pi^0$$

Some recent experiments seem to indicate a fine structure for the ρ^0-meson, and the original broad resonance seems to be due to two closely spaced narrow resonances ρ_1 and ρ_2, each with a width of $\leqslant 30$ MeV (Fig. 1).

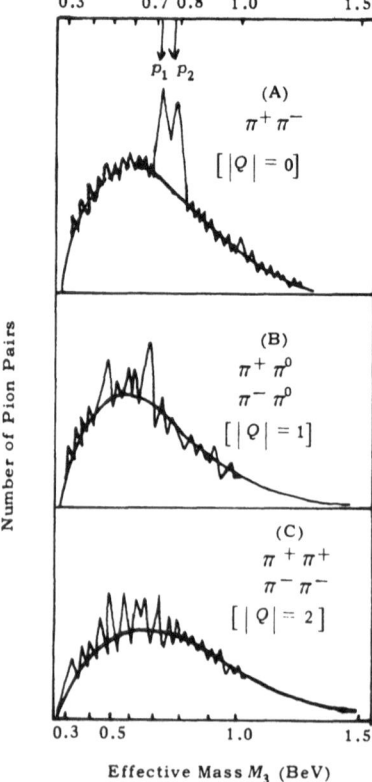

Fig. 1. Fine structure of the ρ^0-meson. (The splitting of the peak occurs only for the $|Q| = 0$ pairs.)

2. THE ω-MESON

Nambu[5] had predicted the existence of an isoscalar vector meson which would explain the isoscalar form factor of the nucleon. The form factor of the nucleon charge distribution is given by

$$F = F^s + F^v$$

(v and s indicate isovector and isoscalar, respectively.)

The isovector part changes its sign going from proton to neutron; the scalar one remains unchanged. Thus, the mean square radius is given by

$$r^2 = (r^v)^2 + (r^s)^2 = 0.8 f \quad \text{for proton}$$
$$r^2 = (r^v)^2 - (r^s)^2 = 0.0 f \quad \text{for neutron}$$

The values 0.8 and 0.0 were obtained from Hofstadter's experiments.[6] Thus, one concludes that $r^v \approx r^s$, and since the dispersion theory gives

$$r^2 = \frac{6}{m_{\text{res}}^2}$$

it follows that

$$m_{\text{res}}^v \approx m_{\text{res}}^s$$

Therefore, the isoscalar particle of Nambu* was expected to have nearly the same mass as the ρ-meson.[7] The mass of the ρ estimated by Frazer and Fulco was 400 to 500 MeV. If so, the ω should decay as

$$\omega \longrightarrow \pi^0 + \gamma$$

and should be found in

$$\gamma + p \longrightarrow \omega + p \longrightarrow \pi^0 + p + \gamma$$

However, no high-energy photon could be detected. m_ρ was later estimated to be 765 MeV, and Sudarshan predicted that

$$m_\rho < m_\omega < m_\rho + m_\pi$$

It was suspected (as seen in the *Introduction*) that it should be possible to observe the ω in

$$p + \bar{p} \longrightarrow N\pi$$

by studying the effective mass distribution of the triplets of pions.

The effective mass distribution of the triplets of pions was computed[8]

---

* A neutral isoscalar meson had been earlier suggested by Teller and Duerr (1956) in connection with the saturation of nucleon forces and nucleon-antinulceon phenomena.[7]

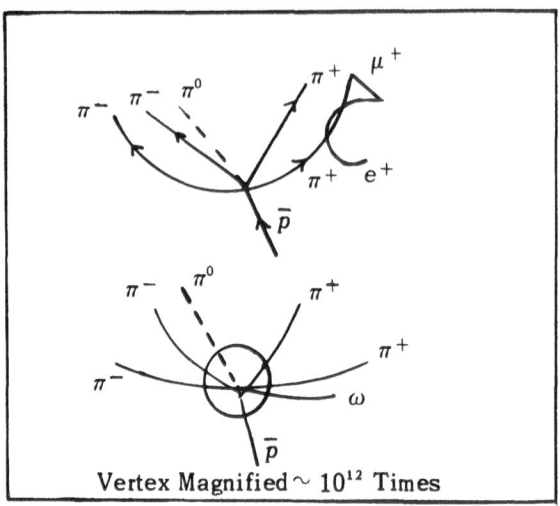

Vertex Magnified $\sim 10^{12}$ Times

Fig. 2. The reaction $p + \bar{p} \rightarrow 2\pi^+ + 2\pi^- + \pi^0$.

using the relation

$$M_3^2 = (E_1 + E_2 + E_3)^2 - (\vec{p}_1 + \vec{p}_2 + \vec{p}_3)^2$$

in the reaction (Fig. 2)

$$p + \bar{p} \rightarrow \pi^+ + \pi^+ + \pi^- + \pi^- + \pi^0$$

The graphs are given in Fig. 3 for triplets with different total charge, namely,

$\lvert Q \rvert = 0$	$\pi^+ \pi^- \pi^0$	four combinations
$\lvert Q \rvert = 1$	$\pi^\pm \pi^\pm \pi^\mp$	four combinations
$\lvert Q \rvert = 2$	$\pi^\pm \pi^\pm \pi^0$	two combinations

In Fig. 3, (A) and (B), which correspond to $\lvert Q \rvert = 1$ and 2, do not have any peak, in contrast to the peak at 787 MeV for the $\lvert Q \rvert = 0$ case.[9] This determines the isospin of ω to be 0 (i.e., $I_\omega = 0$, $M_\omega = 787$ MeV, $\Gamma_{\frac{1}{2}} \leqslant 15$ MeV).

3. DETERMINATION OF THE SPIN OF ω

A $T = 0$ state of a 3π system must be antisymmetric in all pairs

$$\begin{array}{c} 2 \uparrow \\ 1 \downarrow \end{array} \longrightarrow 3$$

Let the 3π system be split into single pion (3) and a dipion (1, 2) systems. Since $T_3 = 1$ and $T_\omega = 0$, we have $T_{12} = 1$. Hence, the

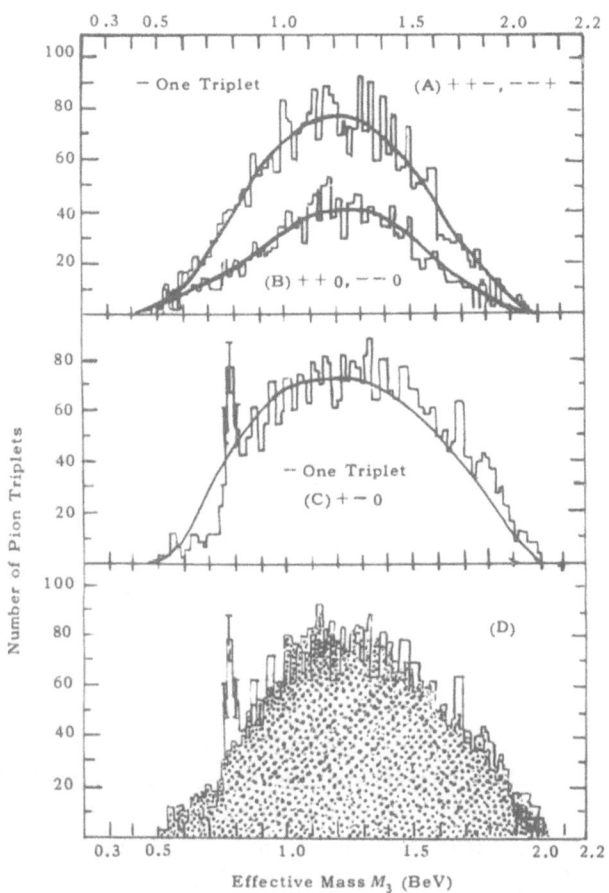

Fig. 3. Number of pion triplets versus the effective mass (M_3) of the triplets for the reaction $\bar{p} + p \rightarrow 2\pi^+ + 2\pi^- + \pi^\circ$. In (D), the combined distributions (A) and (B) (smooth line) are contrasted with distribution (C).

space-wave function of 1, 2 should also be antisymmetric. Let L be the orbital angular momentum of 1, 2 in the dipion rest system. Then the lowest value of L is 1, since it has to be odd. Let l be the orbital angular momentum of particle 3 in the 3π rest system. Then the different spin-parity assignments for ω are those shown in Table I, where spin-2 has been ruled out. These matrix elements are of the simplest possible form. It is then convenient to make a Dalitz plot as shown in Fig. 4.

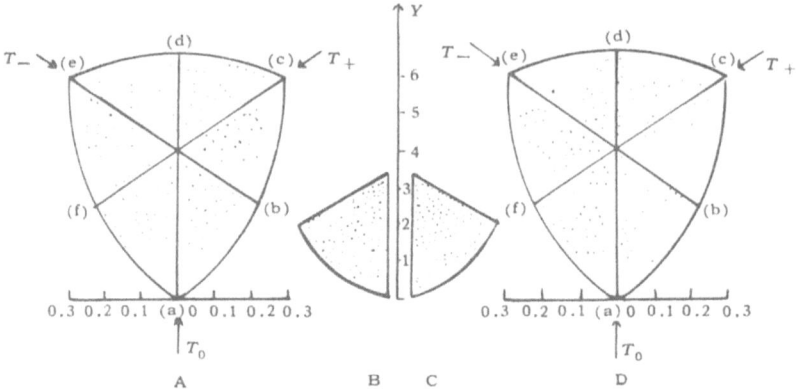

Fig. 4. (A) Dalitz plot of triplets from the control region. (B) Folded control-region plot. (C) Folded peak-region plot. (D) Dalitz plot for triplets in the peak region, 43 % of which are due to ω-mesons.

In the nonrelativistic limit, the conservation of energy and momentum restricts the allowed region to a circle. For the extreme relativistic case it is a triangle. The allowed region for the energy involved in the experiment is shown in the figure as *abcdef*. The size of the figure is proportional to

$$T_1 + T_2 + T_3 = Q = m_\omega - (2m_{\pi^+} + m_{\pi^0})$$

All the three matrix elements are antisymmetric. They vanish whenever one of the pions has its maximum kinetic energy (d, f, b). However, the more striking feature of the peak–region plot is the depopulation when any $\vec{p}_\pi = 0$ (a, c, e). This suggests an angular momentum barrier $(l > 0)$ and constitutes evidence against an axial vector meson. The medians correspond to equal energies of two of the pions. The scalar matrix element will have to vanish on these lines and hence if ω were 0^- there would be depopulation along these lines, which is not observed. The whole boundary of the Dalitz plot corresponds to parallel and antiparallel configurations for the pions, and so the $|(\text{matrix element})|^2$ should vanish at the boundary for a vector meson, which indeed seems to be the case in Fig. 4.

Figure 5 represents the number of events per unit area of the Dalitz plot for the peak region and for the central region versus the distance from the center of the Dalitz plot; the curves expected for 1^+, 1^-, and 0^- mesons are also drawn.[9,10]

The experimental data agree well with the curve predicted by a matrix element of a vector meson (1^-) and not at all with the prediction of 0^- or 1^+ mesons.*

However, these analyses were based on the assignment of G-parity -1 to ω. The Dalitz plot for a 0^- particle with $G = \pm 1$ should be flat, which is excluded by our data.

The effective mass distribution for the neutral ρ^0 shows two peaks at 720 MeV (ρ_1^0) and 780 MeV (ρ_2^0). The latter is very near the ω-meson and suggests the possibility of a G-violating ω decay

$$\omega(G = -1) \longrightarrow \pi^+ + \pi^- + \pi^0 \quad (G = -1) \quad (G \text{ allowed})$$

$$\omega(G = -1) \longrightarrow \pi^+ + \pi^- \qquad\quad (G = +1) \quad (G \text{ forbidden})$$

Feld was the first to indicate the possibility that ω decays to $\pi^+\pi^-$ electromagnetically.

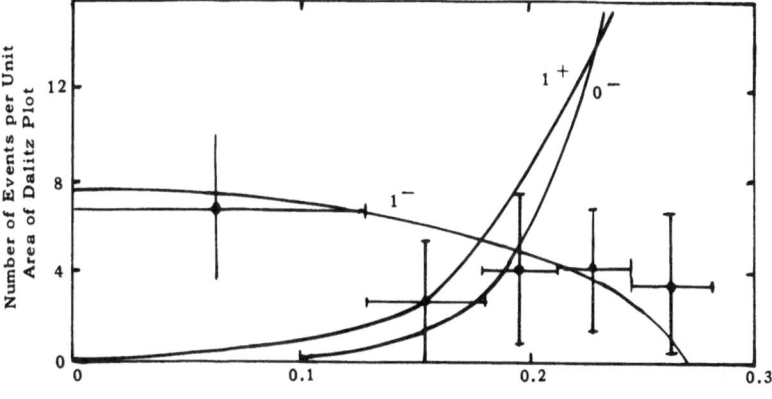

Fig. 5. Number of events per unit area of the Dalitz plot versus the distance from the center of the Dalitz plot.

4. THE η-MESON

Pevsner et al.[11] observed two peaks in the histogram of the effective case of the 3π system in the reactions

$$\pi^+ + d \longrightarrow \pi^+ + \pi^- + \pi^0 + p + p$$

* W. Heisenberg and P. Duerr showed that, taking G-parity into account, only the assignments 0^{-+} and 1^{--} were compatible with the data available on the ω-meson; they pointed out that the observed narrow width of the ω-meson slightly favored 0^{-+}, while the Dalitz plot was in favor of the assignment 1^{--}.

at 550 and 770 MeV, respectively (Fig. 6). The large peak near 770 MeV is clearly identifiable as the ω^0. The other peak is named η-meson. The mass and width calculated are $M_\eta = 546$ MeV and $\Gamma \leqslant 25$ MeV and the quantum numbers tentatively assigned were $1^{-(-)}$ (G-parity [$-$]). Bastien *et al.*[12] have observed the η in the reactions (Fig. 7)

$$K^- + p \longrightarrow \Lambda + \eta^0$$

with

$$\eta^0 \longrightarrow \pi^+ + \pi^- + \pi^0$$

and

$$\eta^0 \longrightarrow \text{neutrals}$$

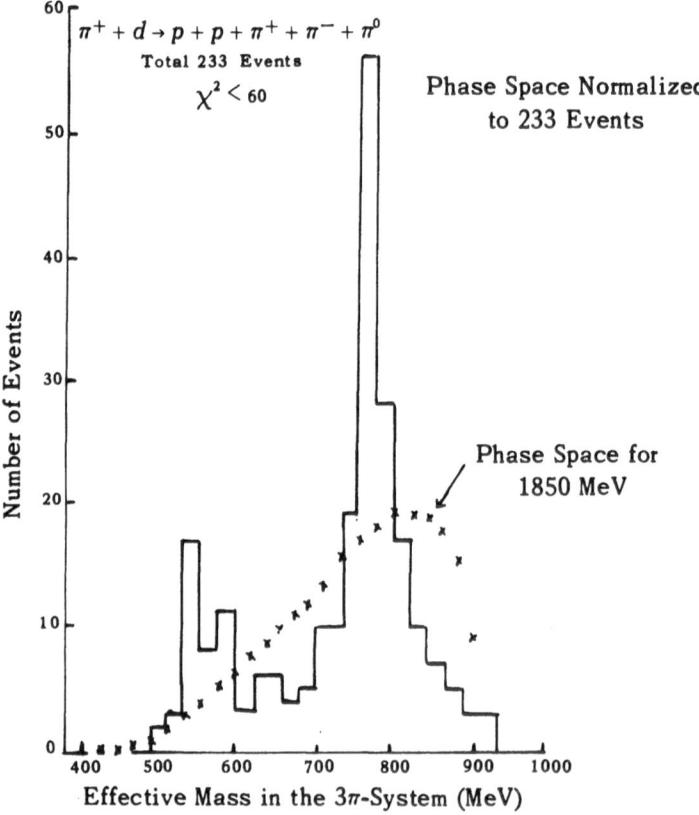

Fig. 6. Histogram of the effective mass of the 3π system in the reaction
$$\pi^+ + d \rightarrow \pi^+ + \pi^- + \pi^0 + p + p.$$

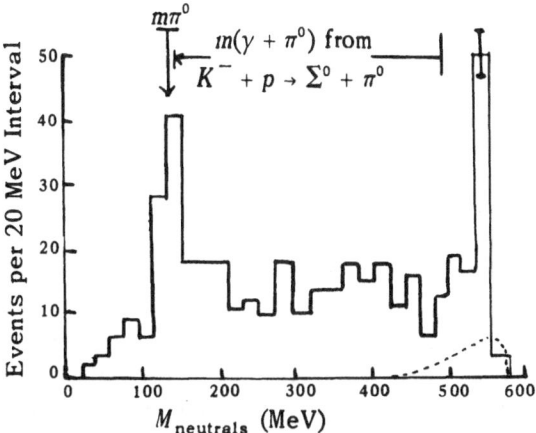

Fig. 7. The reaction $K^- + p \to \Lambda + \eta^0$.

and have deduced the ratio

$$\frac{\eta_{ch}^0}{\eta_{neut}} = 0.31 \pm 0.11$$

They favor 0^{-+} for the η, although 1^{--} is also not ruled out.[†] Pickup, Robinson, and Salant also get a flat Dalitz plot, which probably establishes 0^- for $G = +1$. It is noteworthy that if η is 0^-, 30% of the η-decay seem to violate G-parity, while in the case of ω less than 5% violate G-parity.

5. THE K^*-MESON AND κ-MESON

These resonances are observed in reactions of the type

$$\pi^- + p \to \Lambda + K^+ + \pi^-$$
$$\to \Lambda + K^0 + \pi^0$$

as peaks in the $K\pi$ system, a broad one at 880 MeV with Γ 60 MeV and a very narrow one at 730. The first is called the K^*-meson and the second the κ-meson. Since it is narrow, the κ-resonance is unlikely to be an S-wave resonance and hence κ is expected to be 1^-, while K^* is almost certainly 1^-.

[†] Recently, Shaw and Wong have given reasons for favoring the assignment 0^{--} for the η-meson; they suggest that a small amount of isospin violation distorts the distribution of points on the Dalitz plot from that expected for a particle with $G = -1$.

6. A VIEW TO THE FUTURE: THE SONIC SPARK CHAMBER AND THE MISSING MASS SPECTROMETER

There is no *a priori* reason why there should not exist, apart from the observed 2π and 3π resonance, 4π, 5π, and even 10π resonances, corresponding to higher values of J and higher masses. An instrument is needed which would enable us to observe such massive pionic states, if they exist.

The earlier method of detecting resonances was based on observing the distribution of events as a function of the "effective masses" of groups of n-particles in the final state of the reaction under study (if the resonance being looked for was an n-particle resonance); for example, in the reaction

$$A + B \longrightarrow C + D + E + F$$

one would look for a three-particle resonance in the system (DEF) by plotting the number of events observed against the effective mass $(M_{\text{eff}})_3$ given by

$$(M_{\text{eff}})_3^2 \equiv (E_D + E_E + E_F)^2 - (\vec{p}_D + \vec{p}_E + \vec{p}_F)^2$$

If a resonance existed in the (DEF) system, it would appear as a peak at some value of M_{eff}.

However, this method involves all the uncertainties occuring in the measurement of three-vector momenta. A simpler method is the method of studying the distribution of the "missing mass" $(M_m)_C$ defined by

$$(M_m)_C^2 = (E_{A+B} - E_C)^2 - (\vec{p}_A - \vec{p}_C)^2$$

$(M_m)_C$ is identical to (M_{eff}); the advantage in the missing-mass method is that in only one vector momentum angle is the angle of emission measured. The spark chamber is convenient for the analysis of missing-mass distribution.

A future program consists of the study of the missing-mass distributions in reactions of the type

$$\pi + p \longrightarrow p + n\pi \qquad n = 1, 2, \ldots, 10$$

using the sonic spark chamber. All the resonances in systems with up to ten pions would be located with much less labor than that involved in the earlier bubble-chamber work.

The spark chamber had been earlier used for photographic recording

Table I. Spin-Parity Assignments for ω

"Meson"					Matrix elements	
Type	J	$\vec{l}$ $\vec{L}$	Type	J	Form	Vanishes at
V	1^-	1 1	A	1^+	$E_-\,(p_0 \times p_f) + E_0(p_+ \times p_-)$ $\quad + E_+(p_- \times p_0)$	whole boundary
PS	0^-	1 1	S	0^+	$(E_- - E_0)(E_0 - E_+)(E_+ - E_-)$	a, c, e, b d, f
A	1^+	0 1	V	1^-	$E_-\,(\vec{p}_0 - \vec{p}_+) + E_0(\vec{p}_+ - \vec{p}_-)$ $\quad + E_+(\vec{p}_- - \vec{p}_0)$	b, d, f only

of the track of high-energy particles. A recent innovation[13] in acoustic methods of recording perhaps promises to make the spark chamber one of the fastest devices for use in elementary-particle work. Its advantages are:

1. Angular measurements can be made with it to an accuracy of $\pm$ 1/20 degree.
2. It has a time resolution of the order of 10^{-9} sec; particles appearing earlier or later in time can be distinguished.
3. It is a very fast device, since the necessary information can be deduced by an automatic device without the photographs, etc., required in bubble-chamber work.

A particle passing through a spark chamber triggers a spark between oppositely charged electrodes; the times of flight of the emerging shock wave to four (piezoelectric) microphones locate the position of the spark. The track of the particle is determined from the measured positions of the spark.

I hope that we shall be able to observe, by means of this new instrument, all resonances produced with comparable cross section. Knowledge of the complete mass spectrum of bosons should help in theoretical searches for an underlying law, common to all "elementary particles."

REFERENCES

1. E. Pickup et al., Bull Am. Phys. Soc. 6: 301 (1961).
2. J.A. Anderson et al., Phys. Rev. Letters 6: 365 (1961).
3. A. Erwin et al., Bull. Am. Phys. Soc. 6: 311 (1961).
4. W.R. Frazer and J.R. Fulco, Phys. Rev. Letters 2: 365 (1959).
5. Y. Nambu, Phys. Rev. 106: 1366 (1957).

6. R. Hofstadter *et al.*, *Rev. Mod. Phys.* **30**: 482 (1958); *Phys. Rev.* **110**: 552 (1958) and **111**: 934 (1958).

7. Teller and Duerr, *Phys. Rev.* **101**: 494 (1956).

8. B. Maglić *et al.*, *Phys. Rev. Letters* **7**: 178 (1961).

9. N.H. Xuong and G.R. Lynch, *Phys. Rev. Letters* **7**: 327 (1961).

10. M.L. Stevenson, L.W. Alvarez, B.C. Maglić, and A.H. Rosenfeld, UCRL-9856.

11. Pevsner *et al.*, *Phys. Rev. Letters* **7**: 421 (1961).

12. P.L. Bastien *et al.*, *Phys. Rev. Letters* **8**: 114 (1962).

13. B. Maglić and F. Kirsten, Acoustic Spark Chamber, UCRL-10057 (submitted to *Nucl. Instr. Methods*).

The Higher Resonances in the Pion-Nucleon System

G. TAKEDA

TOHOKU UNIVERSITY
Sendai, Japan

We shall first discuss the experimental situation regarding the higher resonances in the pion–nucleon system and then consider various theoretical approaches that have been made to explain them. The first resonance N_1 in the $J = \frac{3}{2}$, $T = \frac{3}{2}$ state is well understood and we shall not speak about it here.

In addition to the higher resonances N_2 and N_3 in the $T = \frac{1}{2}$ state ($\pi^- - p$ system) and N_4 in the $T = \frac{3}{2}$ state ($\pi^+ - p$ system), two new resonances N_5 and N_6 have been observed recently,[1] the first one in ($\pi^- p$) scattering and the second one in $\pi^+ p$ scattering.

Let us discuss the likely J and T values for the resonances N_5 and N_6, using the experimental data on cross sections. For this purpose let us recall briefly the expressions for the elastic, inelastic, and total cross sections in terms of the phase shifts. If we have an incoming plane wave (corresponding to the incident particles), then in the case of two–particle scattering there will be a difference $(e^{2i\delta} - 1)$ between the incident outgoing spherical wave and outgoing spherical waves which represent the scattered wave. If there is also an absorption of the incident wave, the difference will be $(\eta e^{2i\delta} - 1)$, with $\eta \leq 1$; η will depend on the angular momentum J as is the case for the phase shift δ. The elastic (σ_{el}), inelastic (σ_{inel}), and total (σ_{tot}) cross sections are then given by the following expressions:

$$\sigma_{el} = \sum_J \pi (2J + 1) \lambda^2 \, |1 - \eta_J e^{2i\delta}|^2$$

$$\sigma_{inel} = \sum_J \pi (2J + 1) \lambda^2 (1 - \eta_J^2)$$

$$\sigma_{tot} = \sum_J 2\pi (2J + 1) \lambda^2 (1 - \eta_J \cos 2\delta)$$

If we have a resonance, then $\sigma = \pi/2$, and we have

$$\sigma_{el} = \sum_J \pi(2J+1)\lambda^2(1+\eta_J)^2$$

$$\sigma_{tot} = \sum_J 2\pi(2J+1)\lambda^2(1+\eta_J)$$

so that

$$\frac{\sigma_{el}}{\sigma_{tot}} = \frac{1+\eta_J}{2}$$

$$\sigma_{tot}^{res} = \sum_J 4\pi\lambda^2(2J+1)\frac{\sigma_{el}}{\sigma_{tot}}$$

Now we shall make use of the relation

$$1 \leq \frac{\sigma_{tot}^{res \cdot J}}{2\pi\lambda^2(2J+1)} \leq 2$$

in order to get limits on the values of J for N_5 and N_6. Since the observed bumps in σ_{tot} at the resonance energies are due to σ_{tot}^{res} and only a certain fraction of σ_{tot}^{res} may show up as bumps in σ_{tot}, we obtain from the observed σ_{tot} the following figures:

$$\sigma_{tot}^{res \cdot J} \geq 10 \text{ mb} \qquad \text{for } N_5$$

$$\geq 5 \text{ mb} \qquad \text{for } N_6$$

Since $2\pi\lambda^2 \approx 3$mb for both N_5 and N_6, we get

$$J_5 \geq \tfrac{5}{2} \quad \text{and} \quad J_6 \geq \tfrac{1}{2}$$

If a substantial fraction of $\sigma_{tot}^{res \cdot J}$ shows up in the bump of σ_{tot}, we expect $J_5 = \tfrac{5}{2}$ and $J_6 = \tfrac{1}{2}$ (Fig. 1).

Further knowledge of the spins of these resonances is obtained by drawing the Regge curves. It has been observed that if the squares of the masses of the resonances (in units of the square of the mass of the pion) are plotted against spins, then N and N_3 lie on a straight line making a slope of $\tfrac{1}{50}$ with the x-axis.

Now if we assume that N, N_3, and N_5 are on the same trajectory and the spin of N_5 to be $\tfrac{5}{2}$, we see from Fig. 2 that the line bends, i.e., the spin is coming down with increasing mass. Wigner[2] has shown that for potential scattering $(d\delta)/(dp)$ is greater than $-a$, where δ is the phase shift, p the momentum of the incident particle, and a the distance beyond which the potential vanishes. Sakita and Bincer[3] have extended Wigner's treatment using the dispersion relations technique and found that

$$\frac{d\delta}{dp} > -\left(\frac{3}{8}+n\right)\frac{1}{2\mu}$$

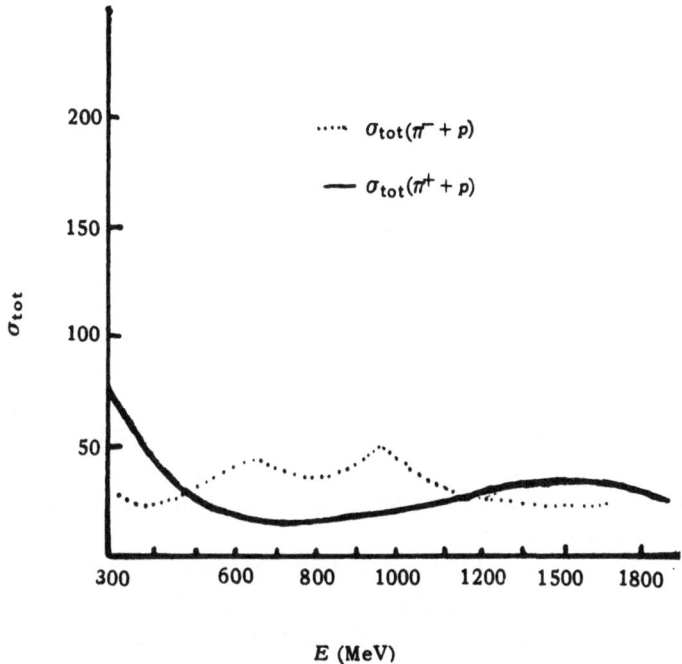

Fig. 1. Total cross sections (σ_{tot}) for π^+-p and π^--p scattering for various values of the kinetic energy E of the pion.

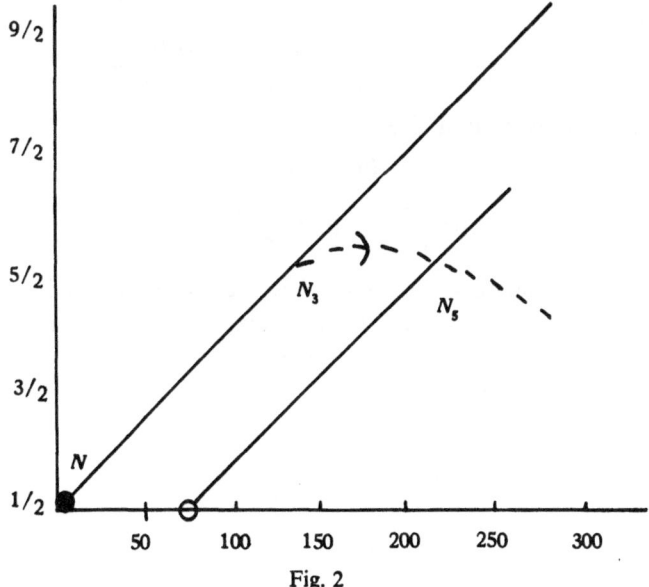

Fig. 2

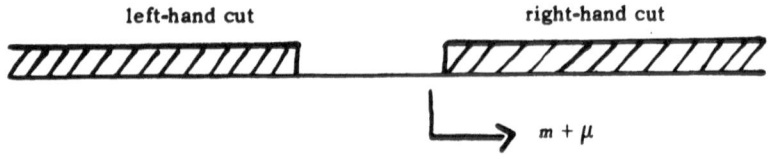

Fig. 3. The cuts in the complex energy plane for $\pi - N$ scattering.

where n is the number of zeros on the left-hand cut in the complex energy plane for π-N scattering and μ the pion mass (Fig. 3).

The width of the resonance is related with $(d\delta)/(dp)$. Putting $n = 0$, we obtain $\Gamma_5 \gtrsim 1.5$ BeV, in disagreement with the experimental value of ~ 0.2 BeV. So this assumption of N_5 being on the NN_3 Regge curve is ruled out.

The second possibility is to allow the N_5 to lie on another Regge trajectory parallel to the NN_3 line. It would then require a resonance in the $P_{1/2}$ state at a lower energy. Some experimental evidence has been available for its existence.

We shall now turn from experimental considerations to what the theorists have done to explain these resonances (N_1 to N_6). In the S-matrix approach, one is interested in the location and strength of the singularities in the complex energy (or momentum transfer) plane. In the complex $w \, (= \sqrt{s}\,)$ plane the position of the poles and branch cuts would appear as shown in Fig. 4. The scattering amplitude can be written as

$$f(w) = \frac{c}{w - m} + \frac{1}{\pi} \int_{m+\mu}^{\infty} \frac{f(w')}{w' - w} \, dw'$$

The unitarity condition gives us

$$\mathrm{Im}\, f = i f^* \rho f$$

$$\mathrm{Im}\left(\frac{1}{f}\right) = \mathrm{Im}\left(\frac{f^*}{f^* f}\right) = -i\rho$$

Here ρ is the phase space factor. For the pion–nucleon intermediate state, ρ is proportional to the center of mass momentum, i.e.,

$$\rho \propto \sqrt{w - (m + \mu)}$$

Now if we write

$$\frac{1}{f} = \frac{1}{A} - i\rho$$

or

$$f = \frac{A}{1 - i\rho A}$$

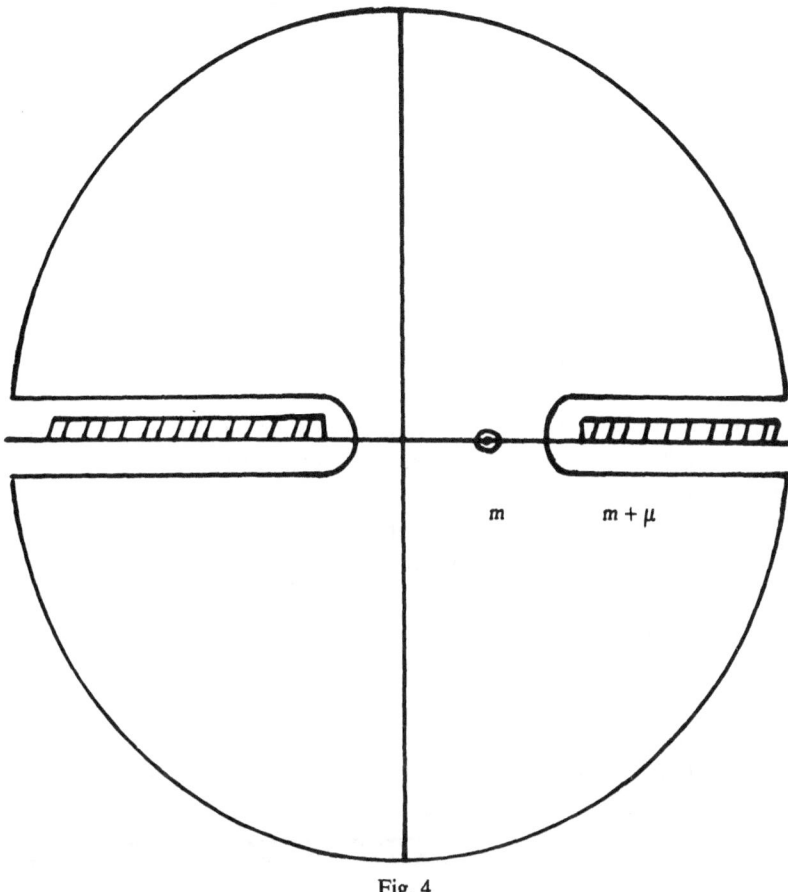

Fig. 4

where A is any smooth function of w, then the unitarity condition on the right-hand cut is taken care of. Near the threshold $m + \mu$, A can be expanded in the form

$$A = A_0 + A_1 q^2 + A_2 q^4 + \cdots$$

Below the threshold, $\rho = i|q|$. Hence if $1 - i\rho\, A \approx 1 + |q|\, A_0 = 0$, which will be the case for $A_0 < 0$, we have a pole of f corresponding to a bound state.

To obtain the singularities corresponding to the resonances we have to continue f analytically through the right-hand cut into the second Riemann sheet. As is well known, poles in these sheets correspond to resonances and branch cuts may correspond to some peculiar anomalies in the cross sections discussed by Tripp[8] and Perez-Mendez.[9] We deform the branch cut, which we had originally taken along the positive real axis, into the second Riemann sheet until it crosses a (possible) pole at m_1 (see Fig. 5). Now in the second sheet ρ approaches $-\rho$, so that the partial wave amplitude in the second sheet is

$$f = \frac{A}{1 + i\rho A}$$

If $1 + i\rho A = 0$ for $w = m_1$, then $1 + i\rho A \propto (w - m_1)$ near $w \approx m_1$. Hence, near the complex pole, we can write the cross section as

$$\sigma \propto |f|^2 \propto \frac{1}{|w - m_1|^2} = \frac{1}{(w - w_R)^2 + (\Gamma/2)^2}$$

where w_R and $\Gamma/2$ are the real and imaginary parts of m_1; $m_1 = w_R - i(\Gamma/2)$. This is of the familiar Breit–Wigner form, so that m_1 indeed represents the position of a resonance.

    The above discussion is a simplified picture of the realistic case. In the case of pion–nucleon scattering, we have to consider, in addition to the graphs corresponding to the poles and cuts in the s-variable (Fig. 6), also the poles and cuts in the u variable (Fig. 7), where

$$u = 2m^2 + 2\mu^2 - s - t$$

t being the momentum transfer.

    The singularities of the partial wave amplitude for pion–nucleon

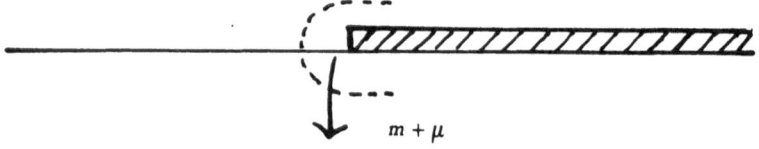

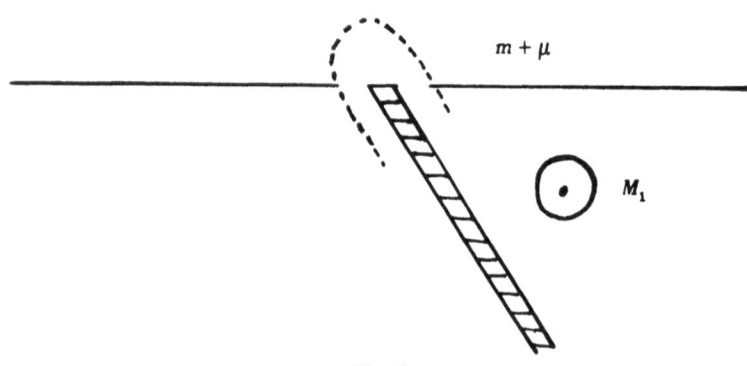

Fig. 5

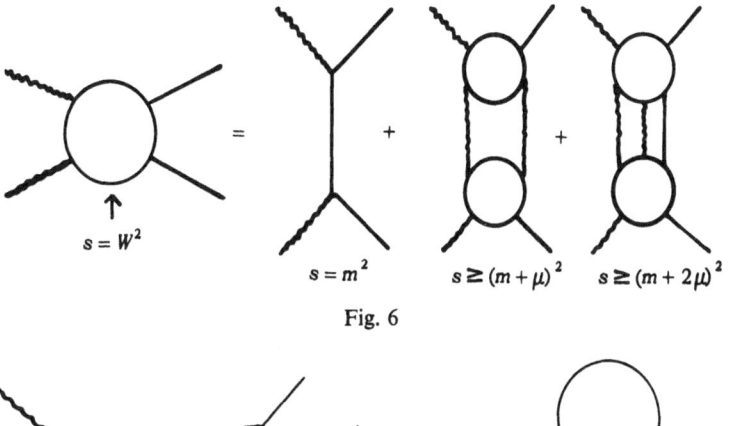

$s = W^2$

$s = m^2$ $s \geq (m + \mu)^2$ $s \geq (m + 2\mu)^2$

Fig. 6

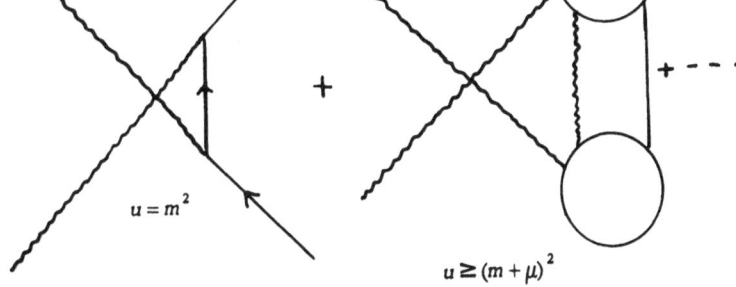

$u = m^2$

$u \geq (m + \mu)^2$

Fig. 7

scattering have been discussed by Frazer and Fulco[4] and by Frautschi and Walecka,[5] among others. In addition to the unitarity cuts from $m + \mu$ to ∞ (the cut from $-m - \mu$ to $-\infty$ can be re-expressed in terms of the positive cut using a symmetry relation of the partial wave amplitudes), there will be other singularities arising from the vanishing of the denominators containing the u and t variables, such as short cuts from $(m^2 + 2\mu^2)^{1/2} \leq w \leq m - (\mu^2/m)$ and $-m + \mu \leq w \leq m - \mu$, the cut along the entire imaginary axis, and a circular cut about the origin having a radius $r = (m^2 - \mu^2)^{1/2}$ (see Fig. 8).

In addition to the above singularities, there may be an additional pole and cut suggested by Peierls.[6] If we consider the process $\pi + N_1 \rightarrow \pi + N_1$, where the resonance N_1 is taken to be an unstable particle, then consideration of Fig. 10a shows us that the isobar N_1 can decay into a pion and a nucleon which subsequently absorbs another pion. This represents a real process and the corresponding pole shows up in the physical region of the momentum and scattering angle. The contribution from this pole shows a resonance-like behavior in pion–nucleon scattering according to the reaction (Fig. 10b).

If we make the partial wave expansion, the Peierls pole leads to a

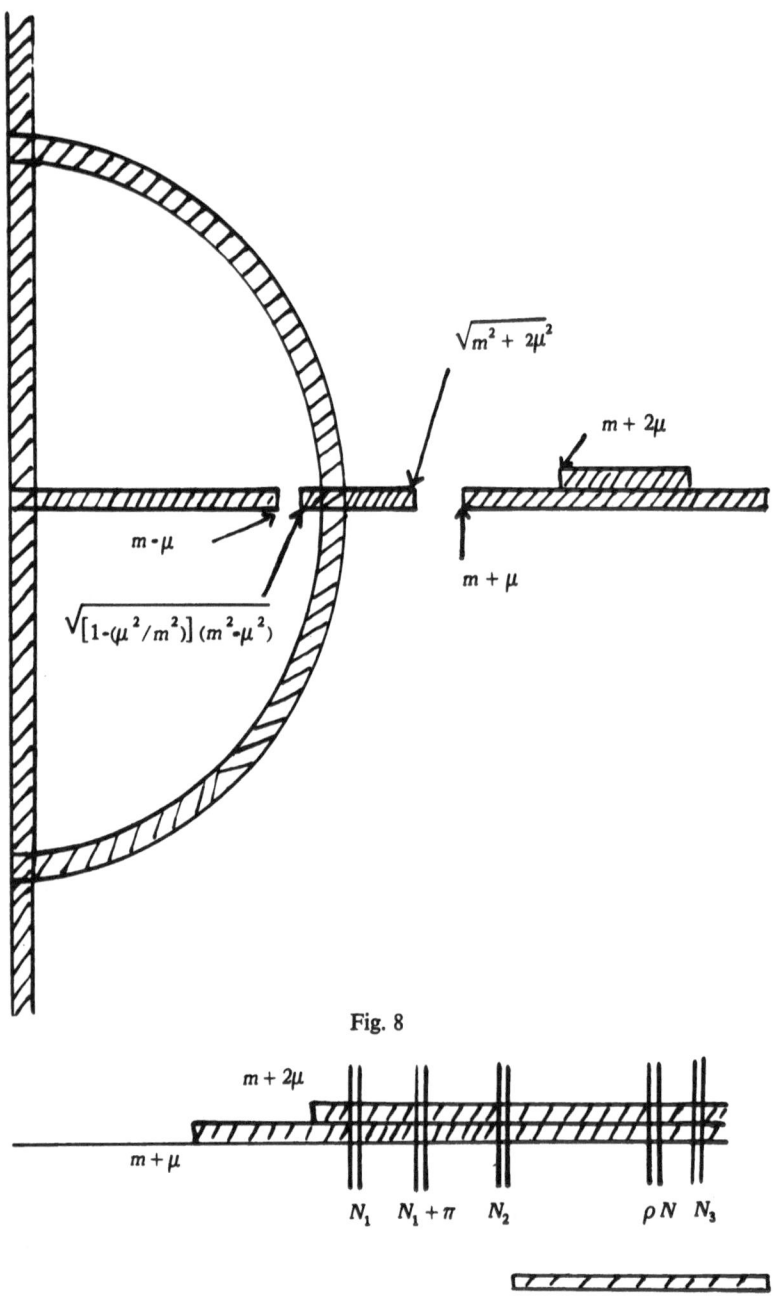

Fig. 8

Fig. 9. The Peierls cut.

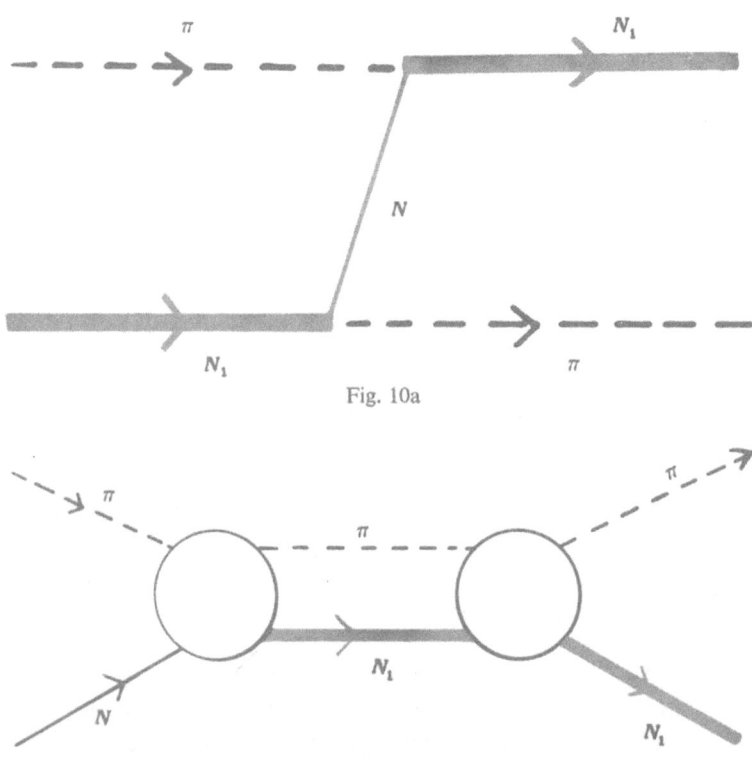

Fig. 10a

Fig. 10b

short cut in the second Riemann sheet of the partial wave amplitude. The location of the cut is rather near to the observed resonance ener-gies of N_2 and N_3. One possible way to understand N_2 and N_3 is to regard the short cut as an approximate pole corresponding to N_2 or N_3. Or one can imagine that the cut produces a pole nearby, which corresponds to N_2 or N_3. Another possible way to understand N_2 and/or N_3 is to associate the occurrence of resonance pole with the presence of a ρN and/or πN_1 channel. For example, if the ρN channel (the threshold for which lies in this region) contributes a negative A to the scattering amplitude, then there is the possibility of a pole.

The assignment of parity and angular momentum for N_2 is not very clear. Recently, the beam strength of the Berkeley bevatron has been increased by a factor of 10 so that formation of the resonant state N_1 can be isolated and a study can be made of the reaction $\pi + N \rightarrow \pi + N_1$ (Fig. 11). The angle χ is measured in the rest system of N_1.

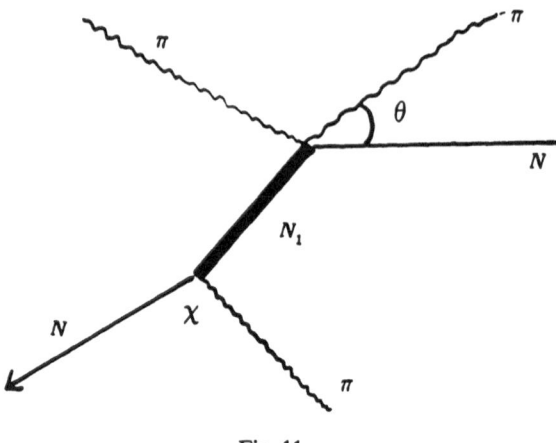

Fig. 11

The angular distribution of πN_1 formation and the decay distribution of N_1 can be measured.

Table I and similar tables for the process $\pi N \rightarrow \rho N$ should be helpful in experimental analysis of πN scatterings near the resonance regions of N_2 and N_3.

We shall now describe briefly the work we[7] have done using perturbation theory which reproduces many of the features of the higher resonances in the pion–nucleon system. The main feature of this approach is the importance of the ρ meson–nucleon intermediate state for πN scattering (Fig. 12a). The $\rho\pi\pi$ interaction Hamiltonian can be written as

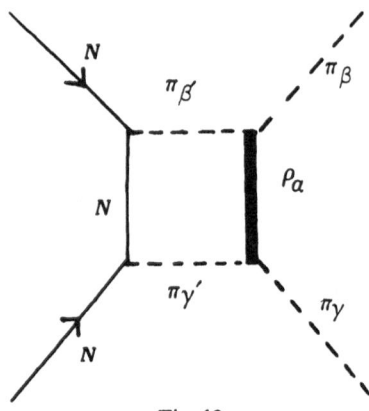

Fig. 12a

Table I

πN	πN_1	$\dfrac{d\sigma}{d\Omega_\theta}$	$\dfrac{(d^2\sigma/d\Omega_\chi d\Omega_\theta)}{(d\sigma/d\Omega_\theta)}$
$s_{1/2}$	$d_{1/2}$	constant	$1 + 3\cos^2\chi$
$p_{1/2}$	$p_{1/2}$	constant	$1 + 3\cos^2\chi$
$d_{3/2}$	$s_{3/2}$	constant	$(1 + 3\cos^2\chi)(1 + 3\cos^2\theta)$ $+\, 9\sin^2\chi\sin^2\theta$
$p_{3/2}$	$p_{3/2}$	$7 - 6\cos^2\theta$	$[(1 + 3\cos^2\chi)(1 + 3\cos^2\theta)$ $+\, 81\sin^2\chi\sin^2\theta]\,(7 - 6\cos^2\theta)^{-1}$
$f_{5/2}$	$p_{5/2}$	$1 + 2\cos^2\theta$	$[(1 + 3\cos^2\chi)(5\cos^4\theta - 2\cos^2\theta + 1)$ $+\, \sin^2\chi\,(-15\cos^4\theta + 14\cos^2\theta + 1)]$ $\times\,(1 + 2\cos^2\theta)^{-1}$

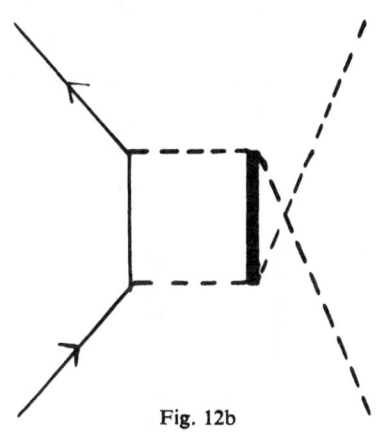

Fig. 12b

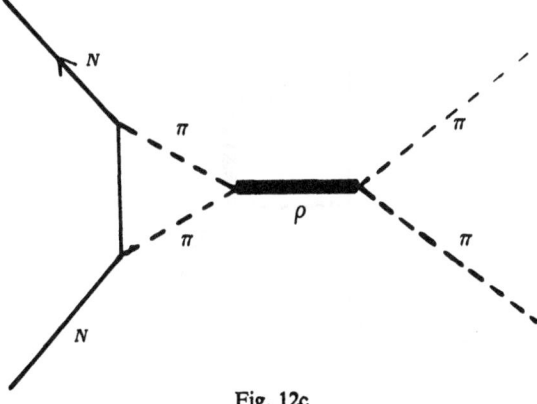

Fig. 12c

$$H_{\rho\pi\pi} = F \sum_{\alpha, \beta, \gamma} \chi_i^\gamma \partial_i \varphi^\alpha \varphi^\beta \epsilon_{\alpha\beta\gamma}$$

where α, β, γ are the isotopic spin indices and χ, φ are the ρ and π field operators. F is the coupling constant and its large value ($F^2/4\pi \approx 5$) will ultimately by found to be responsible for the N_2 and N_3 resonances. The procedure to find F is to calculate the pion–pion scattering amplitude in perturbation theory using the above Hamiltonian and compare it with the Breit–Wigner one-level formula for π–π resonance ρ. Then we obtain $F^2/4\pi \approx 5$.

In the calculation, iteration of the Born term as in Fig. 13 is included. Figure 14 represents the total cross-sectional curves drawn on the basis of numerical calculation at different energies.

The cross section shows a maximum above the value 25 mb. The phase shifts and η corresponding to various values of E_π/μ are given in Table II.

Finally, we will briefly mention a bump at 400 to 500 MeV in the mass distribution of the pions observed in the pion production by pion–nucleon collision. The anomaly was noticed by Tripp[8] and Perez-Mendez[9] independently. It is higher than the ABC anomaly (about

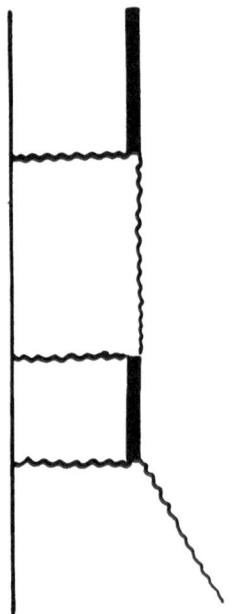

Fig. 13

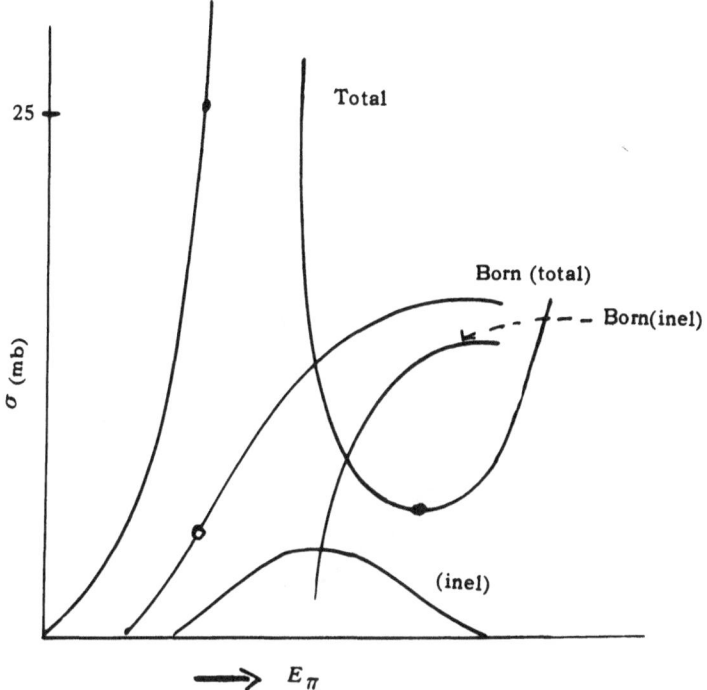

Fig. 14

Table II

	E_π/μ (degrees)		
	3	4	5
δ	38.5	174	154
η	0.94	0.96	0.95

310 MeV) and perhaps swamps it. We can expect it to arise from a complex anomalous singularity (discussed by Landshoff and Treiman[10]) corresponding to a triangle diagram of the kind shown in Fig. 15. This diagram gives rise to a cut in the second Riemann sheet of the two-pion mass plane. The cut ends at a point whose location depends on the total energy of the πN system. Figure 16 shows how the branch point moves with the πN total energy w (M^* is the mass of N_1). For $w \approx M^* + \mu$ where the experiments were performed, the branch point is very near to the physical sheet of the $M_{2\pi}$ plane and may produce the observed bump at $M_{2\pi} \approx 3\mu$.

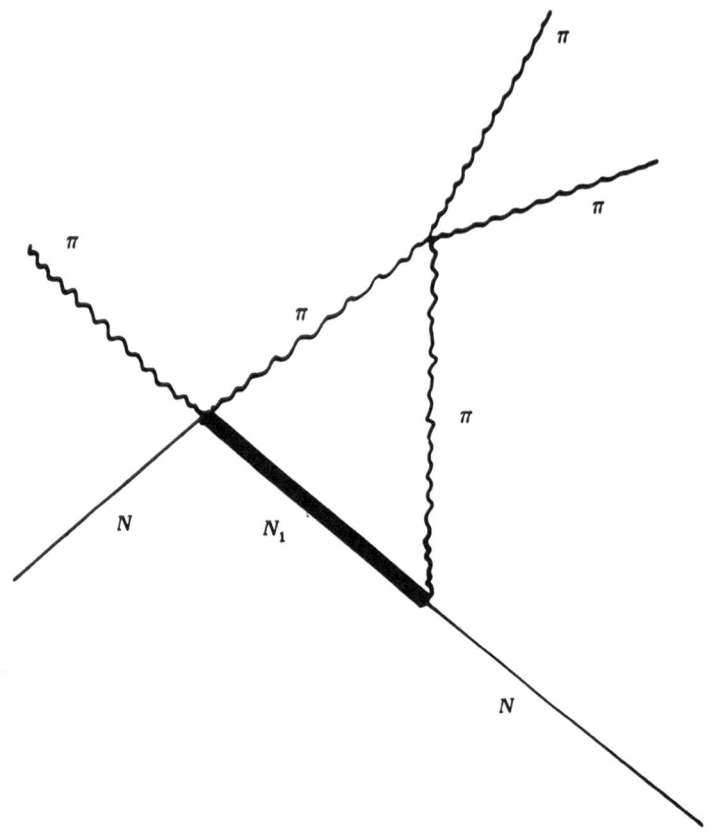

Fig. 15

$M_{2\pi}$-plane

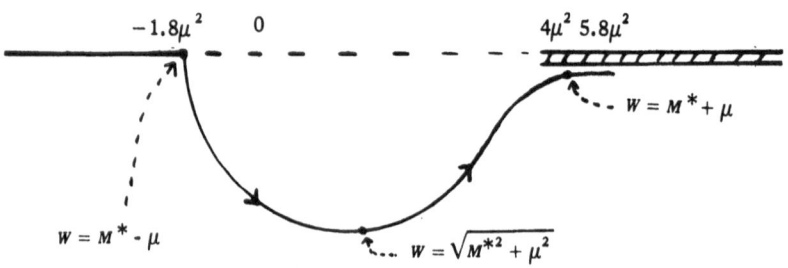

Fig. 16

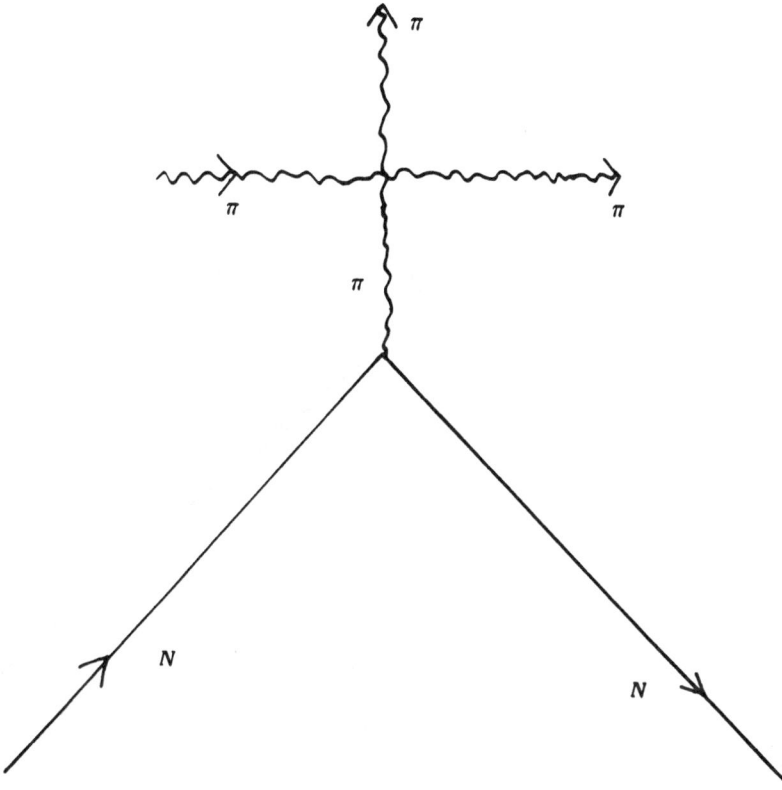

Fig. 17

An alternative explanation might be to consider the one-pion exchange graph (Fig. 17) and diagram (Fig. 18). The corresponding matrix elements are, respectively,

$$M_1 \propto \frac{\vec{\sigma} \cdot \vec{\Delta}}{\Delta^2 + \mu^2} = \frac{|\Delta|}{\Delta^2 + \mu^2} (\sigma_{||} \cos \theta_{\text{lab}} + \sigma_{\perp} \sin \theta_{\text{lab}})$$

Spin-nonflip $\propto \cos \theta_{\text{lab}}$

Spin-flip $\propto \sin \theta_{\text{lab}}$

$$M_2 \propto (\vec{\sigma}_2 \cdot \vec{q}_2^c) - \tfrac{1}{2}(\vec{\sigma} \cdot \vec{q}^c) = \tfrac{2}{3}\sigma_{||} q_2^c - \tfrac{1}{3}\sigma_{\perp} q_{\perp}^c$$

Spin-nonflip $\propto \tfrac{2}{3} \cos \theta_c$

Spin-flip $\propto -\tfrac{1}{3} \sin \theta_c$

Here θ_{lab} and θ_c are the recoil angles of N in the laboratory and center

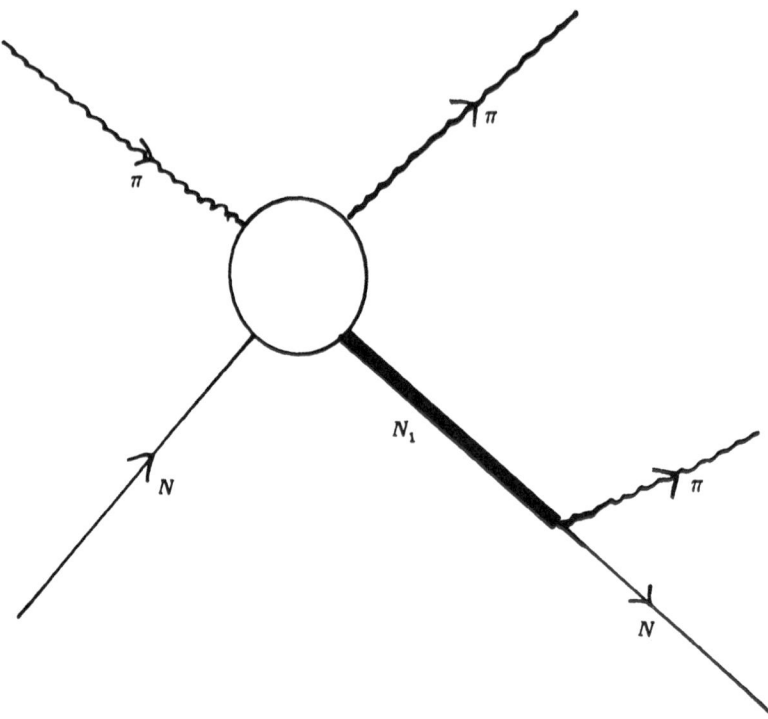

Fig. 18

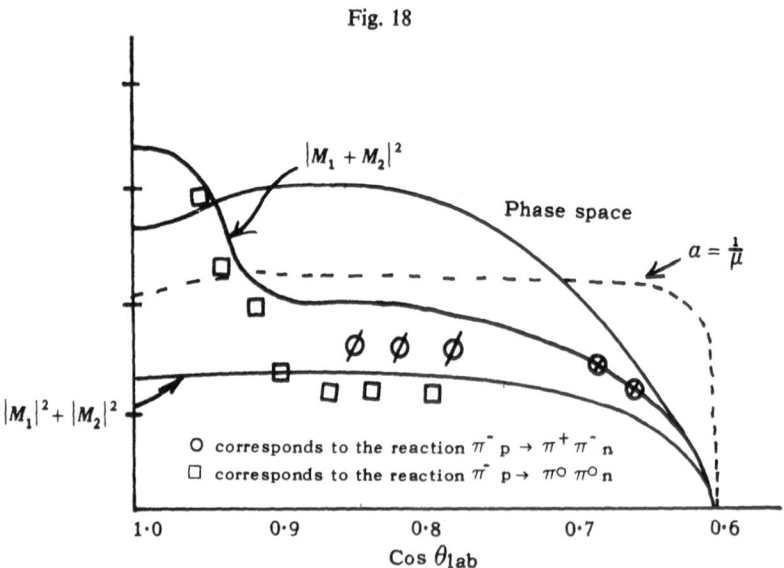

Fig. 19

of mass system, respectively. The angular distribution of recoil N is given by

$$d\sigma \propto |M_1|^2 + |M_2|^2 + \text{interference term}$$

where

$$\text{interference term} \propto \cos\theta_{\text{lab}}\cos\theta_c - \tfrac{1}{2}\sin\theta_{\text{lab}}\sin\theta_c$$

$|M_1|^2 + |M_2|^2$ and $|M_1 + M_2|^2$ are plotted along with the experimental points; thereby the value of $|M_1||M_2|$ is adjusted to give the best fit with experiments (see Fig. 19). The observed bump at $\cos\theta_{\text{lab}} = 1$ corresponds to the bump in the $M_{2\pi}$ distribution. Thus, the interference term seems to be a possible candidate as a source of the Tripp–Perez-Mendez anomaly.

None of the above explanations regarding the anomaly is conclusive. Further study is certainly required.

REFERENCES

1. A.N. Diddens *et al.*, *Phys. Rev. Letters* **10**: 262 (1963).
2. E.P. Winger, *Phys. Rev.* **98**: 145 (1955).
3. A.M. Bincer and B.Sakita, *Phys. Rev.* **129**: 1905 (1963).
4. W.R. Frazer and J. Fulco, *Phys. Rev.* **119**: 1420 (1960).
5. S.C. Frautschi and J.D. Walecka, *Phys. Rev.* **120**: 1486 (1960).
6. R.F. Peierls, *Phys. Rev. Letters* **6**: 641 (1961).
7. K. Itabashi, M. Kato, K. Nakagawa, and G. Takeda, *Progr. Theoret. Phys.* **24**: 529 (1960).
8. J. Kurz, J. Schwarz, and R.D. Tripp, *Phys. Rev.* **130**: 2481 (1963).
9. B.C. Barish, R.J. Kurz, V. Perez-Mendez, and J. Solomon, *Bull. Am. Phys. Soc.* **7**: 280 (1962).
10. P.V. Landshoff and S.B. Treiman, *Phys. Rev.* **127**: 649 (1962).

Author Index

Subject Index

Tentative Data on Strongly Interacting Particles (A. H. Rosenfeld)*

Particle	Established quantum number $I(J^{PG})$	Possible assignment — Quantum number $I(J^{PG})$	Possible assignment — Regge trajectory	Mass (MeV)	Γ (MeV)	Mass² (BeV²)	Dominant decays — Mode	Percent	Q (MeV)	P or P_{max} (MeV/c)
Vacuum (?)	—	$0(2^{++})$	$+\omega_\alpha$	—	—	—				
η	$0(0^{-+})$		$+\omega_\beta$	548	10	0.30	(Even no. of π) K̄, K, etc. Neutrals	75	—	—
							$\pi^+\pi^-\pi^0$	25 ± 4	136	375
ω	$0(1^{-+})$		$-\omega_\gamma$	782	15	0.62	$\pi^+\pi^-\pi^0$	86	368	326
							$\pi^0\gamma$	14 ± 4	647	379
π { π^0	$1(0^{--})$		$-\pi_\beta$	135	0	0.018	$\pi^0 \to 2\gamma$	100	135	67
$\pi^\pm$			$+\pi_\gamma$	140	0	0.02	$\pi^\pm \to \mu\nu$	58	34	30
ρ	$1(1^{-+})$			750	100	0.56	$\pi\pi$	100	471	348
ζ	$1(?)$	$1(0^+)$	$-\pi_\alpha$	560	15	0.31	(p-wave) $\pi\pi$	?	290	245
K { K^0	$\frac{1}{2}(1^-)$		K_β	498	0	0.24	$K^0_1 \to \pi^+\pi^-$	$\frac{2}{3}K_1$	219	206
$K^\pm$	$\frac{1}{2}(1^-)$			494	0		$K^\pm \to \mu\nu$	58	338	236
$K^*_{1/2}(888)$	$\frac{1}{2}(1^-)$		K_γ	888	50	0.78	$K\pi$	100	251($K^0\pi^-$)	283
$K^*_{1/2}(730)$	$\frac{1}{2}(?)$	?	?	730	20	0.53	(p-wave) $K\pi$	?	101($K^-\pi^0$)	161

Particle	$I(J^P)$	Symbol	Mass (MeV)	Width	m^2	Decay mode	%		
$N\left\{\begin{array}{l}n\end{array}\right.$	$\frac{1}{2}(\frac{1}{2}^+)$	N_α	940	0	0.88	$P\bar{e}\nu$	100	0.78	1.2
$\left.\begin{array}{l}p\end{array}\right.$			938			—	—	—	—
$N^*_{1/2}$("900" MeV)	$\frac{1}{2}(\frac{5}{2}^+)$	N_α	1688	100	2.84	$N\pi$	?	610	572
$N^*_{1/2}$("600" MeV)	$\frac{1}{2}(\frac{3}{2}^-)$	N_γ	1512	150	2.28	$N\pi$ (wave)	100	434(π^-p)	450
$N^*_{3/2}$(isobar)	$\frac{3}{2}(\frac{3}{2}^+)$	Δ_δ	1238	100	1.53	$N\pi$ (d-wave)	100	160(π^-p)	233
$N^*_{3/2}$	$\frac{3}{2}(\frac{7}{2}^+)$	Δ_δ	1920	200	3.69	$N\pi$ (p-wave) + others	?	824(π^-p)	722
Λ	$0(\frac{1}{2}^+)$	Λ_α	1115	0	1.24	π^-p	67	38	100
Y_0^*	$0(J\geq\frac{5}{2})$	Λ_α	1815	120	3.29	KN + others	?	383(K^-p)	541
Y_0^*	$0(?)$	Λ_β	1405	50	1.97	$\Sigma\pi$	100	69	144
						$\Lambda 2\pi$		10	69
Y_0^*	$0(\frac{3}{2}^-)$	Λ_γ	1520	15	2.31	$\Sigma\pi$ (d-wave)	60	194	267
						KN (d-wave)	30	88	244
						$\Lambda 2\pi$	10	125	253
Σ^0	$1(\frac{1}{2}^+)$	Σ_α	1191	0	1.42	$\Lambda\gamma$	100	76	74
Σ^-			1196	0	1.42	$n\pi^-$	100	117	192
Σ^+			1189	0	1.42	$n\pi^+$	50	110	185
Y_1^*	$1(\frac{3}{2}^+)$	Σ_δ	1385	50	1.92	$\Lambda\pi$	98	135($\Lambda\pi^0$)	210
Y_1^*	$1(J\geq\frac{3}{2})$		1685(?)	?	2.85	$\Sigma\pi, \Lambda\pi$ + others	2 ± 2(?)	49($\Sigma^-\pi^+$)	119
							?	435	459
$\Xi\left\{\begin{array}{l}\Xi^0\end{array}\right.$	$\frac{1}{2}(\frac{1}{2}^+)$	Ξ_α	1311	0	1.72	$\Lambda\pi^0$	—	61	131
$\left.\begin{array}{l}\Xi^-\end{array}\right.$			1321			$\Lambda\pi^-$	—	66	138
Ξ^*	$\frac{1}{2}(?)$	?	1530	<7	2.34	$\Xi\pi$	100	74	148

*These data are taken from UCRL Report September 1962. The Addendum on the following pages is taken from UCRL 8030, and represents new data developed from 1962 through 1965.

Data on Strongly Interacting Particles (Addendum)

Particle	$I(J^{PG})$	Mass (MeV)	Width (MeV)	Mass² (BeV²)	Partial mode	Fraction (%)	Q (MeV)	P or P_{max} (MeV/c)
						Important decays		
X⁰ or η'	$0(0^{-+})$	958.6 ± 1.6	<4	0.920	$\eta 2\pi$	76±4	131	232
				<0.008	$\pi^+\pi^-\gamma$	24±4	680	459
φ	$0(1^{--})$	1019	3.3±0.6	1.040	$K_1 K_2$	41±6	23	109
					K^+K^-	59±6	32	126
					$\pi\pi$	<8	740	490
f	$0(2^{++})$	1253 ± 20	118±16	1.57	$\pi\pi$	Large	974	611
				0.294	4π	<4	695	547
					$\bar{K}K$	<4	265	386
D	$0(1^{++})$	1286 ± 6	40±10	1.65	$\bar{K}K\pi$		154	303
				0.105				
E	$0(-)$	1420 ± 10	60±10	2.02	$K^*\bar{K}$	Large	35	151
				0.17	$K\bar{K}\pi$		293	430
f'	$0(2^{++})$	1500	80	2.25	$K_1 K_1$		505	561
				0.24	$K\bar{K}^*(890)$		111	274
A_1	$1(1^{+-})$	1072 ± 8	125	1.150	$\rho\pi$	~100	167	231
				0.27	$\bar{K}K$	<5		
B	$1(\geq 1^{++})$	1220	125±17	1.488	$\omega\pi$	~100	298	339
				0.31	$\pi\pi$	<30		
					$K\bar{K}$	<10		
					4π	<50	662	528
A_2	$1(2^{+-})$	1324 ± 9	90±10		$\rho\pi$	~91	419	426
					$\bar{K}K$	~5.5±1.5	333	439
					$\eta\pi$	3.6±3.0	636	537

C	$\leq\frac{3}{2}(-)$	1215 ±15	60±10	1.476, 0.145	$K\rho$	?	−44	<0	
					$K^*\pi$	?	184	253	
K^*	$\frac{1}{2}(2^+)$	1405 ±8	95±11	1.988, 0.27	$K\pi$	?	775	610	
$N^*_{1/2}$	$\frac{1}{2}(\frac{1}{2}^+)$	1480	→	Existence not yet well established					
$N^*_{1/2}$	$\frac{1}{2}(\frac{7}{2}^-)$	2190	~200	4.80, 0.88	πN	<10	1112	888	
					ΛK	Seen	577	710	
$N^*_{1/2}$	$\frac{1}{2}(\frac{9}{2}^+)$	2650	~200	7.00, 1.06	πN		1567	1151	
					ηN		1158	1090	
$N^*_{3/2}$	$\frac{3}{2}(\frac{9}{2}^-)$	2360	~200	5.58, 0.95	πN	~15	1282	988	
$N^*_{3/2}$	$\frac{3}{2}(\frac{11}{2}^+)$	2825	260	7.98, 1.47	πN		1747	1252	
Y^*_1	$1(\frac{5}{2}^-)$	1762 ±17	75±7	3.11, 0.26	$\bar{K}N$	~60	328	496	
					$\Lambda\pi$	~16	510	517	
					$\Sigma\pi$	≤3	436	467	
					$Y^*_1(1385)+\pi$	~10	243	318	
					$Y^*_0(1520)+\pi$	~10	107	188	
Y^*_1	$1(\frac{7}{2}^+)$	2065	~160	4.26, 0.74	$\bar{K}N$	~0.35	632	728	
					$\Lambda\pi$		819	726	
Ξ^*	$\frac{1}{2}(\frac{3}{2}^-)$	1816 ±3	16±4	3.27, 0.06	$\Xi^*\pi$	~25	149	232	
					$\Lambda\bar{K}$	~65	205	394	
					$\Xi\pi$	<5	362	413	
Ξ^*	$\frac{1}{2}(\frac{5}{2}^+)$	1933 ±16	140±35	3.74, 0.54	$\Xi\pi\pi$	<5	222	312	
					$\Xi\pi$		473	501	
Ω^-	$0(\frac{3}{2}^+)$	1675 ±3		2.81	$\Xi\pi$	?	221	296	
					$\Lambda\bar{K}$	?	66	216